Nelson
Technology Activity Manual
Third edition **Basil Slynko**

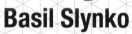
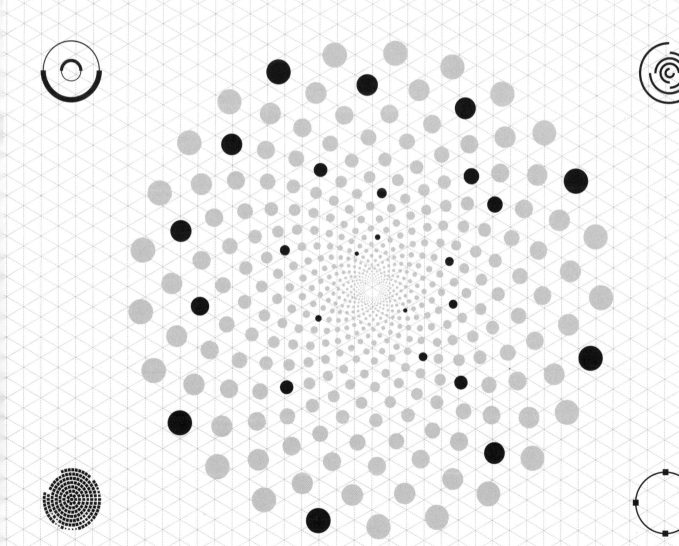

Nelson Technology Activity Manual
3rd Edition
Basil Slynko

Associate publisher: Siobhan Moran
Publisher: Sam Bonwick
Editor: Georgia O'Connor
Project editor: Georgia O'Connor
Text designer: Nadia Backovic
Permissions researcher: Debbie Gallagher
Project designer: James Steer
Illustrator: Paul Lennon
Production controller: Karen Young

Any URLs contained in this publication were checked for currency during the production process. Note, however, that the publisher cannot vouch for the ongoing currency of URLs.

For product Information and technology assistance,
in Australia call **1300 790 853**;
in New Zealand call **0800 449 725**

For permission to use material from this text or product, please email **aust.permissions@cengage.com**

ISBN 978 0 17 043990 9

Cengage Learning Australia
Level 7, 80 Dorcas Street
South Melbourne, Victoria Australia 3205

Cengage Learning New Zealand
Unit 4B Rosedale Office Park
331 Rosedale Road, Albany, North Shore 0632, NZ

For learning solutions, visit **cengage.com.au**

Printed in China by 1010 Printing International Limited
4 5 6 7 24

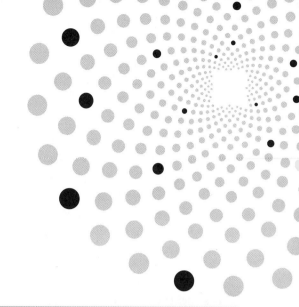

Contents

Introduction

To the teacher ...

Nelson Technology Activity Manual (third edition) is a design thinking handbook: it uses a problem-solving approach to foster technological literacy.

Nelson Technology Activity Manual provides a wide range of activities that cover over 30 major topics in design and technologies including design, systems, materials, tools, processes and safety. The order of the topics can be changed to suit individual teaching styles and syllabus requirements.

This activity manual concludes with nine 'Enrichment' activities designed to challenge your students. Each 'Challenge' is an opportunity for students to display their knowledge and understanding of the principles of design and technologies they have learnt. These challenges encourage students to use higher levels of cognitive thinking. You might wish to consider a collaborative small-group approach when attempting some of the challenges.

You will find an Enrichment icon **e** at the end of some worksheet. This is a cross-reference indicating that there is an extension activity for that particular worksheet.

At the end of each worksheet there is a performance scale, and a place for your signature and the date. The assessment criteria are explained in 'To the student ...' on this page.

Nelson Technology Activity Manual can be used independently or in conjunction with other texts such as *Nelson Introducing Technology* (fourth edition).

NelsonNet

Sample answers for all activities are available on NelsonNet, but alternative answers are possible in a number of cases.

Complimentary access to NelsonNet is available to teachers who use *Nelson Introducing Technology* (fourth edition) as a core resource in their classroom. Contact your education consultant for access codes and conditions.

To the student ...

What is technology?

> **Technology** (tek-NOL-e-ji), the branch of knowledge that deals with science and engineering, or its practices, as applied to industry; applied science (*Macquarie Dictionary, Seventh Edition*, 2017)

Nelson Technology Activity Manual (third edition) will help you gain a general awareness of technology and develop design thinking. Completing the 31 worksheets will provide you with the basic knowledge necessary for mastering design and technologies, not only at school, but also well beyond your school career.

The nine enrichment worksheets are an opportunity for you to apply the basics of design and technologies that you have learnt in new settings. You are encouraged to complete each 'Challenge'. Your teacher may suggest a small-group approach when attempting some of the challenges.

At the end of each worksheet there is a performance scale. Your teacher will assess you as being at one of the five grades. They are:

- **A**, which means that you have completed the exercise extremely well
- **B**, which means that you have completed the exercise with minimal errors
- **C**, which means that you have completed the exercise to an acceptable standard
- **D**, which means that you have completed the exercise with a larger number of errors than expected
- **E**, which means that you should revise, using your textbook, and then resubmit your work.

Acknowledgements

Nelson Cengage wishes to gratefully acknowledge the contribution of E. Mazurkiewicz to the first and second editions of *Nelson Technology Activity Manual*.

Basil Slynko would like to thank the following people and businesses for advice and assistance: Alison Gierke, Mark Pierce and Prototype Design, Ikea Brisbane, Lindsay Wright, Bill Thomas, Andrew Mackay, Rex Faldt, David Thurlow, Amanda Slynko and Doug Turnball.

Name.. Date..

1.1 Technology

These activities will help you to:

▶ gain knowledge of how people use technology to solve problems

▶ gain knowledge of different steps in technology

▶ know the meanings of various terms.

Technology is people using **resources** such as

| Ideas | Information | Skills | Materials | Tools and equipment | Energy |

and a **problem-solving process** such as

| **Investigating** Coming up with a general idea of how to solve the problem and considering the impacts on people and the environment. | ▶ | **Research and planning** Creating a detailed plan of how to produce the solution | ▶ | **Producing and implementing** Making the solution using resources such as tools, materials and skills | ▶ | **Testing and evaluating** Testing the solution to see if it is satisfactory, and assessing its effects on people and the environment |

to produce | a satisfactory solution – product, process, environment or system – to a problem. |

Q1 These images show how people have used technology to solve problems in the world around them. Write a sentence or two about how well each solution solves a problem. (What are the advantages and disadvantages?)

Image A: Indigenous Australians fishing in the 1800s

Image B: Traditional bamboo scaffolding in China today

Advantage: ....

....

Disadvantage: ....

....

Advantage: ....

....

Disadvantage: ....

....

Q2 In Australia today, what solutions do we have for the technology problems shown in images A and B on the first page of this worksheet?

A: ....

B: ....

Q3 In your opinion, which solution is better in each case: the modern solution, or the traditional solution shown in the images? Discuss your views with other students in your group or class.

Q4 Choose either image A or B and complete the comparison chart below.

The problem I am investigating is ....

Past solution:	**Present solution:** ....
Past materials:	Present materials:
Impact then:	Impact now:
Advantages/disadvantages:	Advantages/disadvantages:

In modern industrial societies there are many different branches of technology – for example, communication technology and food technology.

Q5 Name two other branches of technology in Australia and their areas of focus (that is, the main types of problems they are designed to overcome).

1 .... Focus ....

....

....

2 .... Focus ....

....

....

Some solutions have been discovered by accident, known as serendipity. One example is Post It Notes®.

Q6 Through research and/or class discussion, record three other examples of 'serendipity' and list the solutions below.

1 .... 2 ....

3 ....

e WORKSHEET 9.1

Teacher's signature   Date

....

A	B	C	D	E

Name.. Date..

Design principles

These activities will help you to:

▶ identify elements and principles of design

▶ understand how the design of an item relates to its function

▶ identify examples of good and bad design.

A designer needs to consider the **elements** and **principles** of design. These give an item its visual appeal. The elements of design are points, lines, shapes, forms, colours, tones and textures.

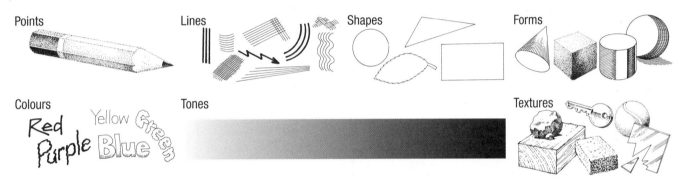

Points   Lines   Shapes   Forms

Colours   Tones   Textures

The principles of design include balance, unity, proportion, contrast, emphasis, movement and rhythm.

Q1 Collect photographs of different versions of a single item – such as a range of microwave ovens or mobile telephones – from magazines or catalogues, or searching online.

Select the two designs that you like the most. Then label some of the elements of design and principles of design you can see in the photographs. Attach the labelled photographs below.

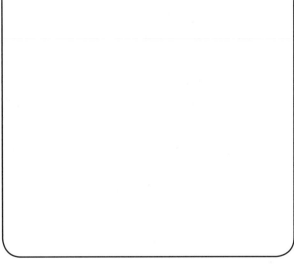

The visual appeal of an item is important. However, designers also have to ensure that the features of their design suit the function of the item.

Q2 Look at these designs for a toy wooden hammer and a gas stove. Consider their functions, then circle the good and bad features of each item and state your reason(s) for circling these features.

Q3 Would you buy these items? Write a sentence about each one to explain your answer.

Toy hammer: Yes ☐ No ☐

....

....

Gas stove: Yes ☐ No ☐

....

....

Some items may have a secondary purpose as well as a primary purpose. For example, the primary purpose of a pair of sunglasses is eye protection. Sunglasses can also be an item of fashion, or they can be used to identify a group in society – this is their secondary purpose.

Q4 Do the above two items have a secondary purpose? (If 'yes', what is it?)

Toy hammer: Yes ☐ No ☐ Secondary purpose: ....

....

Gas stove: Yes ☐ No ☐ Secondary purpose ....

Q5 Look at the two items below, and think about their function and appearance. Then consider how these items could be improved by being redesigned. (Note: Redesign only the control panel for the microwave oven.)

Q6 What modifications could you make to each product to improve its function and appearance? Use the grids below to sketch your solutions. Then briefly describe your reasons for the modifications.

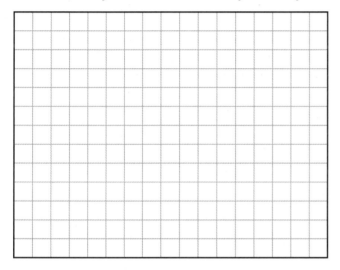
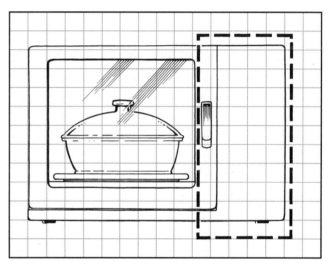

My reasons: ....

....

....

....

....

My reasons: ....

....

....

....

....

Q7 Now show your modifications to some of your classmates and discuss your designs with them. Record their comments below.

Favourable comments: ....

....

....

Unfavourable comments: ....

....

....

Teacher's signature   Date

....

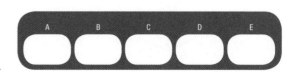

WORKSHEET

2.2

Rendering and colour

These activities will help you to:

▶ develop rendering techniques

▶ understand the use of colour.

Designers use rendering to add realism to their drawings. Rendering involves using **lines or dots** to create tones (shading) and texture. The lines and dots can be spaced or angled differently to suggest a variety of **finishes** such as shiny surfaces, textured surfaces and wood grains.

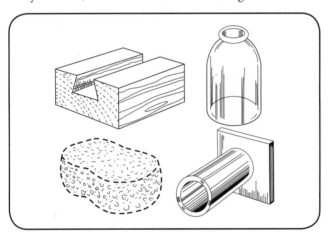

Lines and dots can also be used to create a **three-dimensional effect** by suggesting tones and shadows.

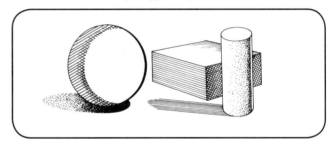

Different weights of lines can also be used to emphasise particular features.

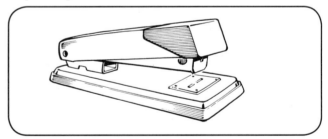

Q1 Render each object below to suggest its appropriate texture.

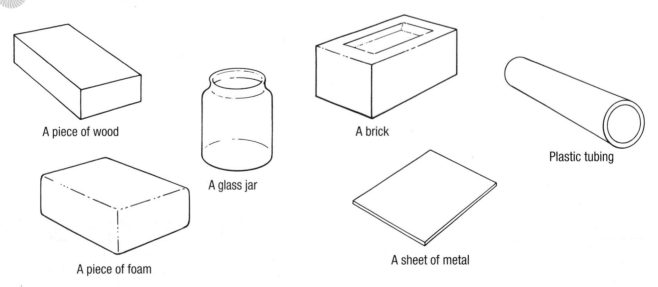

A piece of wood

A glass jar

A brick

Plastic tubing

A piece of foam

A sheet of metal

The elements of design are points, lines, shapes, forms, colours, tones and textures. These are the basic building blocks of design.

Colour is very much part of our daily life. Colours are used to create moods or feelings, or to attract our attention. We see colour in fashionable clothing, in commercial packaging and in safety colour coding. Designers use colour to add to the overall effect of a design and its aesthetic appeal.

A colour wheel shows a range of different colours and their relationship to each other. It has three stages:

First stage: primary colours – red, blue and yellow. These colours cannot be made by mixing any other colours.

Second stage: secondary colours – a mixture of primary colours. For example, red and yellow → orange.

Third stage: tertiary colours – primary and secondary colours mixed together. For example, red and violet → purple (plum); blue and green → deep green (aqua).

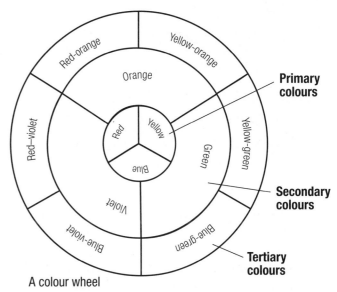

A colour wheel

Q2 Use colour pencils or other suitable media to colour in the colour wheel.

Colours can be used to create contrast or harmony.

Colours opposite each other on the colour wheel (e.g. red and green) give **contrast**. These are known as **complementary** colours.

Q3 List two other complementary colours: .... and ....

Harmony is achieved when you combine colours that are close to each other on the colour wheel. These are known as **analogous** colours. Examples are red and orange.

Q4 Record two other analogous colours: .... and ....

Colours are used to create moods or feelings.

Q5 Which colour(s) would you use to create the following effects?

Peaceful: ....

....

Restless: ....

....

Hot: ....

....

Cool: ....

....

Q6 Collect some pictures of products that use colour to achieve complementary effects or analogous schemes. Then attach them below for future reference.

WORKSHEET
9.3

Teacher's signature   Date

....

WORKSHEET
2.3 Modelling

These activities will help you to:

▶ analyse a manufactured object

▶ develop modelling skills.

Every day we use many different objects, each one designed for a specific purpose – for example, a shampoo bottle. A great deal can be learnt by studying these design solutions.

Q1 Select a manufactured object and analyse it in the space below. Your analysis should state why you think it is designed the way it is. (You may use a sketch or picture of the selected object to help you to explain your comments.)

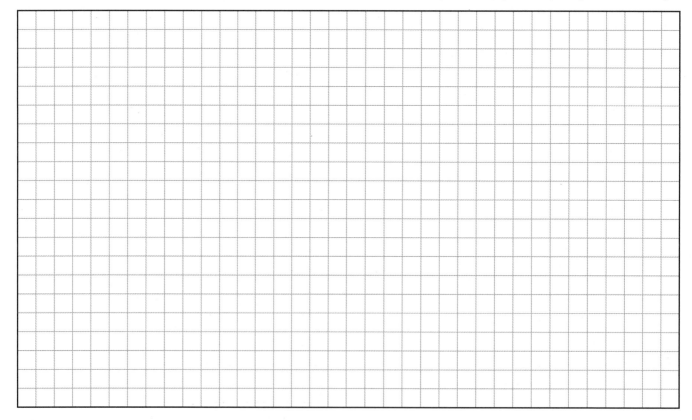

Anthropometry is the science that deals with human body measurements for different genders and different age groups.

The study of people in relation to their home, work or leisure environment (including space, lighting and layout) is known as **ergonomics**.

Anthropometry and ergonomics are two very important factors that must be considered when designing anything to meet people's needs.

9780170439909 © 2019 Basil Slynko

Q2 What anthropometric data and ergonomic factors would you need to consider if you were asked to design a table and chairs for preschool children?

....

....

....

....

....

....

Modelling is a quick way to help you visualise and test a design before you make the item. The model or mock-up could be full size, or it could be scaled down if it is too large to make.

You can use inexpensive materials such as paper, cardboard, plastic sheeting, glue and staples to produce mock-ups. Alternatively, more realistic models can be made from foam, wood and metal, applying appropriate finishes.

Q3 Collect some examples of unusual and interesting packaging. The primary function of packaging is usually to hold and protect the contents. Does the packaging you have collected have any secondary functions? If so, note them here.

A scaled-down model of a building

Secondary functions of packaging: ....

....

Q4 You have been asked by a client to make a full-size mock-up of a package to hold golf balls. The package must be able to hold three golf balls and must also advertise the product inside it. Using a grid sheet, make your mock-up (the diameter of a golf ball is 40 millimetres).

Q5 Now show your mock-up to other students in the class and discuss your solution with them. Record the comments you receive.

Favourable: ....

....

....

Unfavourable: ....

....

....

....

Teacher's signature   Date

....

These activities will help you to:

▶ read a working drawing

▶ compile a list of materials needed to make an item

▶ develop a procedure.

Here is an order for timber. It shows the common system used when ordering stock (standard-sized) timber.

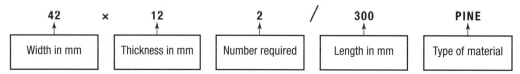

42	×	12		2	/	300		PINE
Width in mm		Thickness in mm		Number required		Length in mm		Type of material

Note: You might need to specify the width, thickness, length, type of material and cross-sectional shape (for example, round tube or hexagonal bar) for some materials.

Sheet material, whether it is metal, plastic or a timber product (for example, plywood) is ordered in this way:

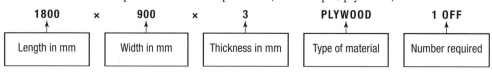

1800	×	900	×	3		PLYWOOD		1 OFF
Length in mm		Width in mm		Thickness in mm		Type of material		Number required

Q1 This is a pictorial drawing of a wooden box for carrying gardening tools. Use the above systems of ordering to list the sizes and quantities of the various pieces needed to make it. Begin with the measurements.

Long side: .... × .... / ....

Short side: .... × .... / ....

Handle: .... × .... / ....

Base: .... × .... × ....

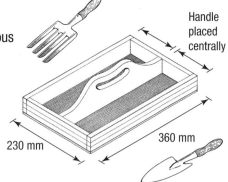

Handle placed centrally

230 mm

360 mm

Material sizes:
• Box 42 mm × 12 mm
• Handle 93 mm × 12 mm
• Base 4 mm plywood

Note: Timber is generally sold in preset sizes that increase by 300 millimetres.

The shortest length sold is 0.9 metre or 900 millimetres long.

Q2 Complete the timber order that you would need to place for the box.

42 × 12 .... / ....;   93 × 12 .... / ....

To make an item, you should develop and use a **procedure**.

Q3 How would you assemble the pieces of the box in this diagram? Place numbers in the circles to show the order in which you would do it.

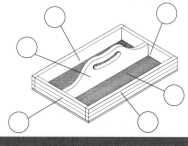

A   B   C   D   E

Teacher's signature   Date

....

9780170439909 © 2019 Basil Slynko

WORKSHEET
2.5
A design process

These activities will help you to:

▶ gain knowledge of a design process

▶ design a functional item

▶ create a logical work sequence

▶ compile a list of materials needed to produce an item

▶ calculate the cost of an item.

Design thinking is the process of planning an outcome and is usually achieved using a problem-solving process such as:

Identifying and defining	**Researching and plannning**	**Producing and implementing**	**Testing and evaluating**
Coming up with a general idea of how to solve the problem and considering the impacts on people and the environment.	Creating a detailed plan of how to produce the solution safely.	Making the solution safely using resources such as tools, materials and skills.	Appraising the solution to see if it is satisfactory and assessing its effects on people and the environment.

The above four step problem-solving process can be expanded into more basic steps, as shown in the diagram on the right.

You will notice that there is a starting point and a suggested flow pattern. The design process can be **cyclical** (continuous). Sometimes the order might not be cyclical, as illustrated in the diagram by the broken arrowlines. You may need to return to an earlier stage(s) to solve the problem.

Start
1 The design brief
2 Analysing the problem
3 Developing your ideas
4 Choosing your final solution
5 Making working drawing(s)
6 Manufacturing the item
7 Testing and evaluation

A design process: a basic approach in seven steps

1 **The design brief** A clear statement of the general problem or need that lists the requirements. This is often referred to as the *design specifications*.

You have just been given 20 classic DVDs of an older relative's favourite groups, movies and concerts but you have no way of storing them. You want to make a container that will hold them.

Q1 Write a design brief in your own words.

....

....

....

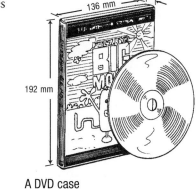

14 mm
136 mm
192 mm

A DVD case

2   Analysing the problem Writing down the information you need to consider. Some possible headings are:

Time limit: How urgent is the need? How much time do you have to make it?

Materials: What is suitable? What is available? What is the cost?

Function: Why is the item needed?

Ergonomics: How efficiently will people be able to use the item? Will its size and shape meet their needs in the best way possible?

Appearance: How will the item look? Style: traditional/modern? What finish could you give it?

Personal safety: Could it injure users or bystanders?

Design consequences: Is there any risk that making the item will harm the community or the environment? Does it conserve natural resources as much as possible?

Construction: Do you have the skills, knowledge and equipment needed to make it?

Q2   Decide on the points that you would need to consider when making a container to hold 20 of the DVD cases shown in the drawing on the previous page. Then write them in this table (examples have been given for ergonomics).

Points to consider	
Time limit	....
Materials	....
Function	....
Ergonomics	Weight of materials if box is to be portable. Ease of removal of contents (space between DVDs) ....
Appearance	....
Personal safety	....
Design consequences	....
Construction	....
Other	....

9780170439909 © 2019 Basil Slynko

Do you have enough information to solve the design problem? If so, continue to step 3 of the design process. If not, seek more information from sources such as your teacher, the library, museums and experts in industry.

3   **Developing your ideas** At this stage, you should record *all* your ideas. Use words and/or pictures.

• Make initial sketches. These could be pictorial drawings (such as isometric or cabinet oblique drawings), or orthogonal drawings (front, top and side views). Remember to keep your sketches in proportion.

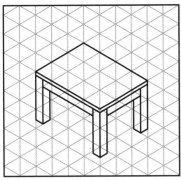

Isometric sketch

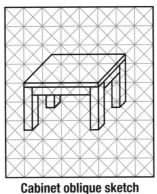

Cabinet oblique sketch

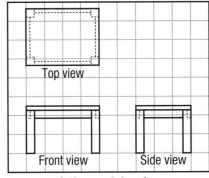

Orthogonal drawing

• Note any features that you might find difficult to sketch (for example, the type of material, size and type of fastener).

Q3   Now develop your ideas on these grid sheets.

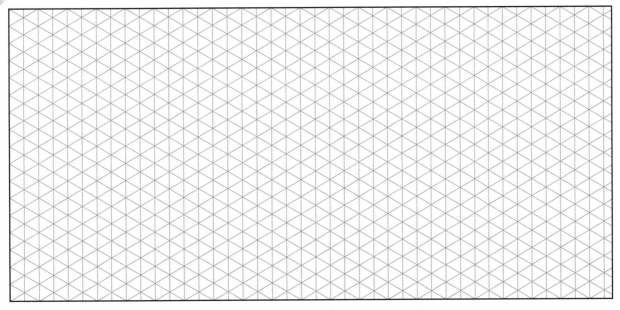

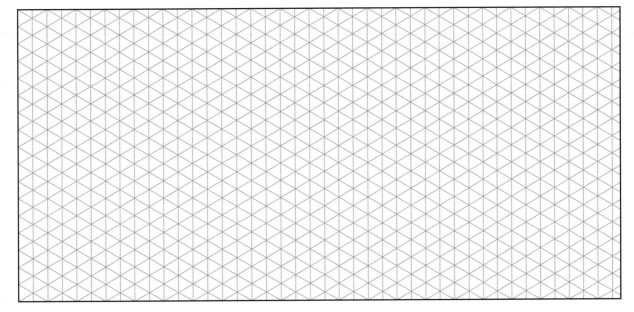

4   **Choosing your final solution** Your first idea might not be the best solution. You might need to combine several ideas to come up with a final solution.

- Study your initial sketches and the features you have noted.

- Choose your final solution, and record all the requirements.

- This record of requirements is called a **final specification**.

5   **Making working drawing(s)** Make a detailed drawing (usually orthogonal) of the components, and write down your specifications. The working drawing should include your procedure (the steps and tools/equipment needed to make the item) as well as a list of materials and the costs involved.

Q4   Now complete your working drawing on the next page (using an appropriate scale), and write your material list in the table provided here. You might also like to use the next page to sketch and render a pictorial drawing of your final solution.

Material list

Component*	Material	Size	Number required	Cost

*Include all items required, such as adhesives, finishes and fasteners.   **Total cost: $**

Note: In industry, common headings are: Number, Description, Quantity, Length, Width, Thickness, Material, Cost and Notes (which could include things like manufacturing methods).

6   **Manufacturing the item** Complete the item using tools and the processes of forming, separating and combining. (Note: Tools and processes are covered in Worksheets 6.1 – 6.4 and Worksheets 7.1 – 7.7.)

7   **Testing and evaluation** You need to test to see whether or not your design has solved the problem.

At this stage, your product is a **prototype**. The procedure of testing a prototype is called evaluation. In industry, any modifications to a product are made at the prototype stage, before tooling up for mass production.

Group activity

Q5   Now discuss your design solution with a classmate and record the comments you receive here.

....

....

....

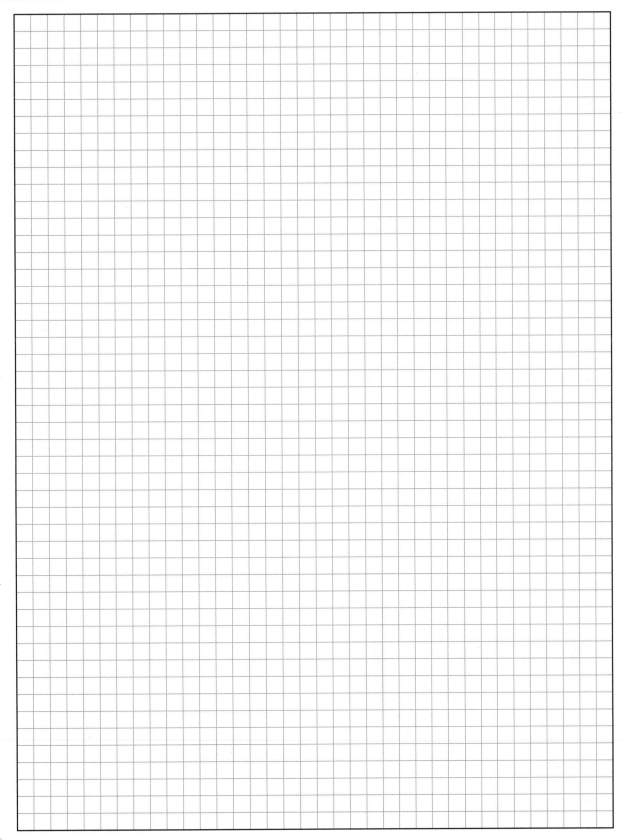

Q6 What are your teacher's comments?

....

....

WORKSHEET 9.3

Teacher's signature   Date

....

WORKSHEET
3.1 Manufacturing

These activities will help you to:

▶ become aware of the world of manufacturing

▶ know the various stages of manufacturing

▶ apply production methods in the workshop.

Manufacturing is the process of making an item that satisfies people's needs or wants. There are a number of stages in the manufacturing process:

1 Researching consumer demand
Before manufacturing a new product, the company must carry out research to find out what consumers want and what price they would pay for it.

2 Developing the product
A team of designers and engineers make prototypes and evaluate them to create a product that not only looks good but also works efficiently.

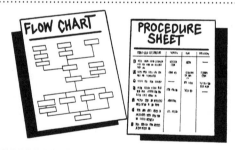

3 Production planning
Every stage of production from raw materials to packaging is planned to avoid delays, maintain quality and keep production costs low.

4 Tooling up for production
Machines and other equipment are obtained and installed according to the production plan so that production can begin.

5 Production (with quality control)
Finished products are made and packaged. Their quality is maintained through a system of checking or testing.

6 Marketing and distribution
The packaged products are warehoused, marketed and transported to purchasers. The selling price should cover all costs and allow for a profit.

This flow chart shows the basic steps of a production procedure in manufacturing:

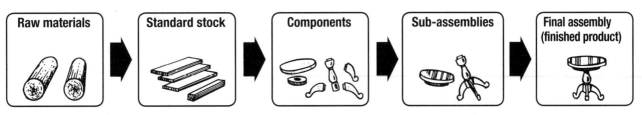

| Raw materials | Standard stock | Components | Sub-assemblies | Final assembly (finished product) |

First, **raw materials** may have to be converted into **standard stock**.

Components (parts) are then produced from the stock.

Different components are put together to make a **sub-assembly**.

Sub-assemblies are combined with other sub-assemblies to form a **finished product**.

Note that manufacturing firms need not carry out all of these processes themselves. They can use stock, components or sub-assemblies that have been made by other firms, known as specialist suppliers.

The task of the manufacturer is to combine the stock, components or sub-assemblies to form the finished product.

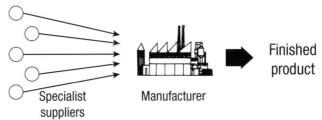

Specialist suppliers

Manufacturer

Finished product

Q1 Study the following production procedure for a pencil and identify each step. Then draw a detailed flow chart in the space provided on the next page of this worksheet to show these steps.

Birth of a pencil

Pencil wood comes from Cedar trees grown in North America. The trees are felled and the wood is taken by sea to the factory.

Graphite is mined in Sri Lanka and Korea.

The rock is crushed into a black powder.

The black paste is squeezed through a small hole into leads, cut into lengths and baked in a hot oven.

The graphite powder and the china clay are mixed together to form a black paste.

White china clay is dug from the ground.

The baked leads are put between two pieces of cedar wood and glued together.

The wood with the leads is then cut into pencils and painted with bright colours. The points are sharpened and the pencils stamped with the name and degree of hardness of the lead.

The pencils are finished.

Flow chart: pencil production

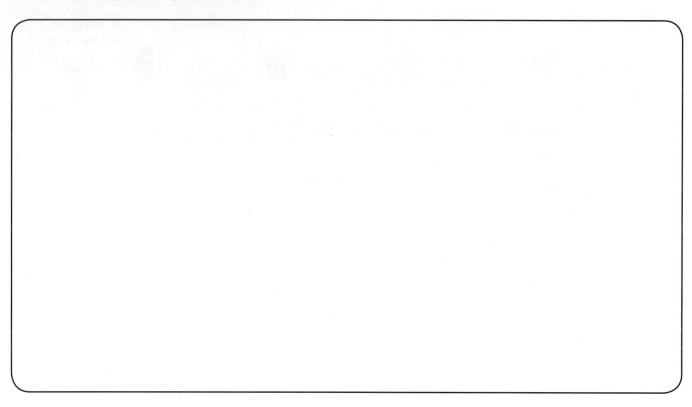

Line production (or mass production, or continuous production, or flow production) is generally used to mass-produce large quantities of a single item. Workers and machines are positioned at workstations along an assembly line. The item is built up as it moves from one workstation to the next. Each worker performs one task in the process.

An alternative way of manufacturing items has been developed at the Volvo automobile assembly plant in Uddevalla, Sweden.

The **Uddevalla** system is based on teamwork. The team members have to assemble the entire product at one location. They are also responsible for their own quality control. Each member of the team is trained to perform a number of tasks – and some members can perform *all* the tasks. A replacement worker is therefore not needed when a worker is absent.

Q2 Complete the comparison chart below.

Line production	Uddevalla production

Q3 Manufacturing crossword

Across

1   A ... sheet lists steps and tools for a production sequence

4   An abbreviation for 'computer-aided manufacturing'

6   'One-off' or ... production

7   In ... production, many of the same item are produced

8   CNC stands for '... numerically controlled'

10   A word that refers to the process of drawing a preliminary design or plan

13   A device used to check whether or not the size of an item is acceptable

17   Has six stages, from research to marketing and distribution

20   An outline for marking or cutting shapes

21   Checking and testing a product is ... control

Down

2   ... materials are converted into standard stock

3   To collect information

4   Individual parts of a manufactured item

5   A chart used to show the sequence of production steps

6   An abbreviation for 'computer-aided design'

9   The permitted difference in size of a manufactured item

11   The setting up of machines and equipment before production begins (two words)

12   The use of machines and systems rather than people is called ...

14   Components are put together to make a ...

15   CIM stands for 'computer-... manufacturing'

16   The actual process(es) by which an item is made

18   Selling goods to consumers

19   A ... is a fixed number of an item that is made (two words)

Name.. Date..

The manufacturing process consists of a number of problem-solving stages, as shown on the first page of this worksheet.

Stage 3 of this process is **production planning**. Every aspect of production from raw materials to packaging is planned using flow charts and procedure sheets to avoid delays, maintain quality, and keep production costs low.

Examples of a production flow chart and a procedure sheet for manufacturing a peg board are shown below.

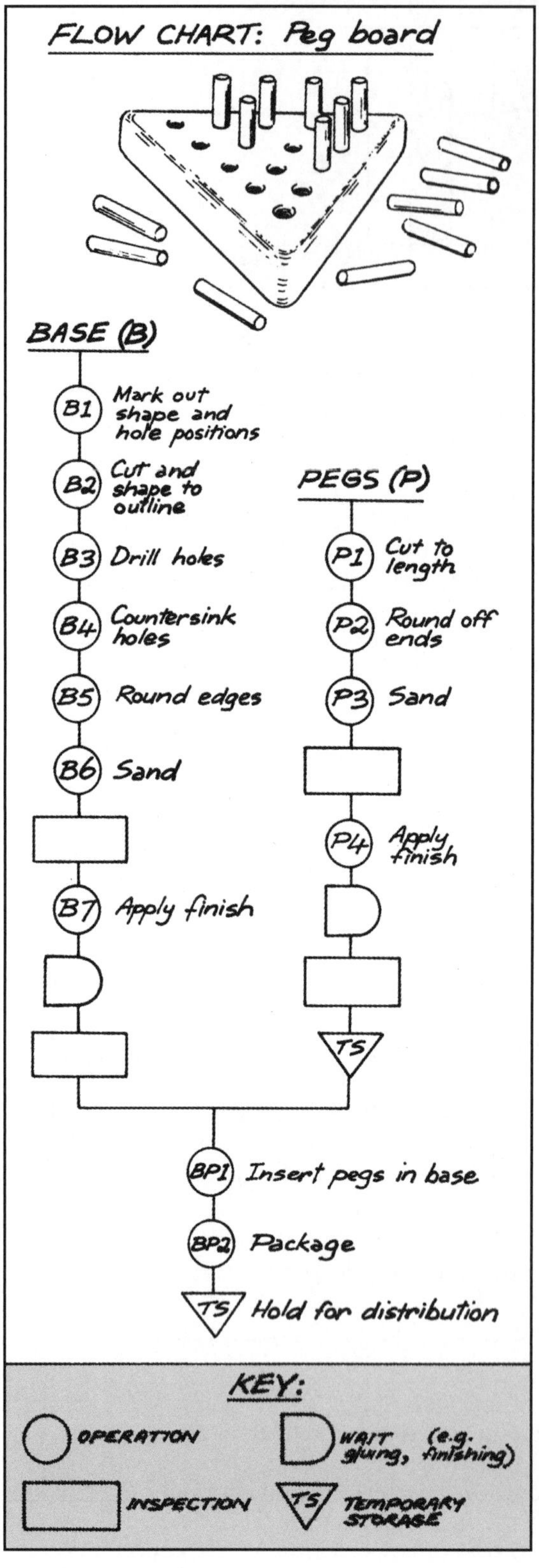

FLOW CHART: Peg board
BASE (B)
B1 Mark out shape and hole positions
B2 Cut and shape to outline
B3 Drill holes
B4 Countersink holes
B5 Round edges
B6 Sand
B7 Apply finish
PEGS (P)
P1 Cut to length
P2 Round off ends
P3 Sand
P4 Apply finish
TS
BP1 Insert pegs in base
BP2 Package
TS Hold for distribution
KEY:
OPERATION
WAIT (e.g. gluing, finishing)
INSPECTION
TS TEMPORARY STORAGE

PROCEDURE SHEET: Peg board

Manufacturing sequence	Operation	Tools and equipment *
	BASE:	
B1	Mark out shape and hole positions	Template
B2	Cut and shape to outline	Hand saw/jig saw/band saw
B3	Drill holes	Bench drill/portable electric drill in drill stand
B4	Countersink holes	Rose countersink with bench drill/portable electric drill in drill stand/brace/hand drill
B5	Round edges	Surform/files/rasps
B6	Sand	Belt sander/orbital sander/hand sand
B7	Apply finish	Brush/spray
	PEGS:	
P1	Cut to length	Hand saw and jig
P2	Round off ends	Disc sander/linishall sander/hand sand
P3	Sand	Hand sand
P4	Apply finish	Spray/brush – colour?
	ASSEMBLY:	
BP1	Insert pegs in base	
BP2	Package	

* You could add another column to record additional information and notes.

To complete the following activity you will need to refer to the working drawing and material list for the DVD container you designed in Worksheet 2.5.

Q4 Prepare a production plan for the mass production of the DVD container that you designed in Worksheet 2.5. Use the peg board example on the opposite page as a guide. You have to:

1 draw up a flow chart for the production sequence

2 draw up a procedure sheet.

Flow chart: DVD box	**Procedure sheet: DVD box**

Teacher's signature   Date

....

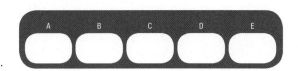

WORKSHEET

3.2 Structures

These activities will help you to:

▶ gain knowledge of different kinds of structure

▶ understand the different kinds of forces that act on a structure

▶ learn some ways in which structures can be made more rigid.

People have made many different kinds of structures that carry loads and withstand other forces. Examples of man-made structures are those that:

* provide shelter

* carry loads

* control the environment.

Structures can be classified as:

* frame structures, which consist of separate pieces of material called members. Members are joined together to form basic shapes such as rectangles and triangles.

* shell structures, which consist of an outside skin that carries the load. They have no frame.

Frame structure   Shell structure

Q1 Which manufactured structures can you identify in the scene on the next page of this worksheet? Classify at least four of them according to the headings in the table below. Also state whether the structure is a frame structure (F) or a shell structure (S).

Structures that provide shelter	Structures that carry loads	Structures that control the environment

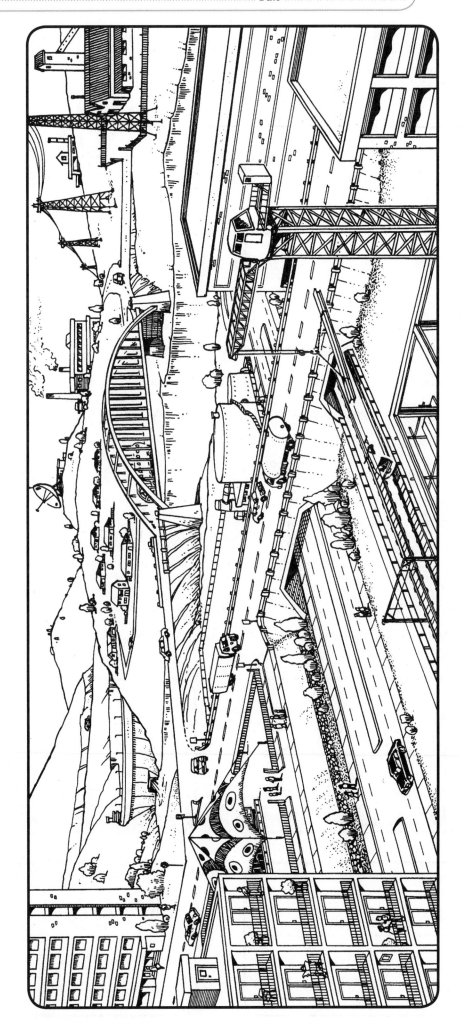

There are two basic kinds of load on any structure: **static** (a load that is still) and **dynamic** (a moving load).

Q2 Identify the load in each case below.

Load ....

Load ....

Structural failure

Q3 Look at the drawing of structural failure above. Then complete this sentence:

Structural failure has occurred because the structure has failed to withstand a .... load.

Q4 Use the word bank to identify the different kinds of force that act on a structure.

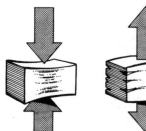

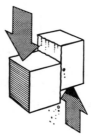
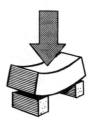

WORD BANK
Bending
Compression
Shear
Tension
Torsion

....

Q5 Study these structures and then identify the forces that act on them (circled). Write your answers on the broken lines provided.

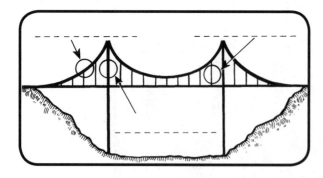

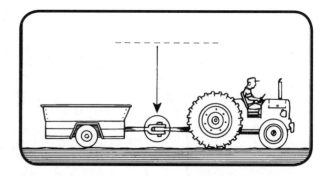

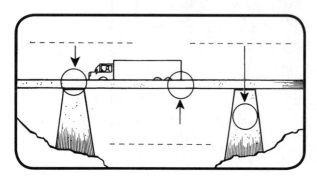

A structure can be made more rigid by building it out of triangular shapes. This is known as **triangulation**.

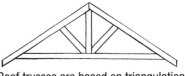

Q6 Sketch where you would attach a bracing member in each case below, given that wood is strong in compression and steel is strong in tension.

Roof trusses are based on triangulation.

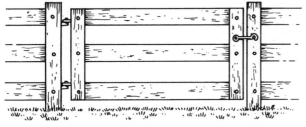

A wooden gate

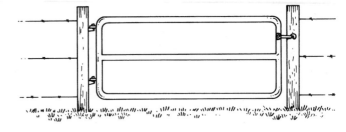

A steel gate

Q7 Show how you can make each structure below more rigid: the sign and the cube box.

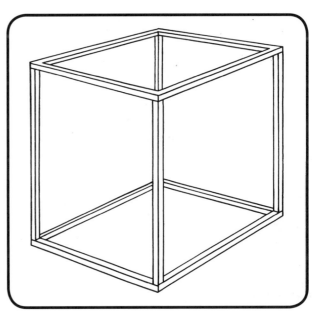

Q8 Sketch some of the ways in which this thin, flat sheet of material can be made more rigid.

Teacher's signature   Date

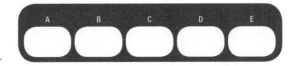

....

WORKSHEET

3.3 Control systems

These activities will help you to:

▶ gain knowledge of control systems

▶ understand how control systems work.

People have developed control systems so that tasks can be done in an organised way. For such systems to work efficiently, there must be:

- an **input** – the system has to be activated
- a **process** – the input is processed
- an **output** – a result is produced.

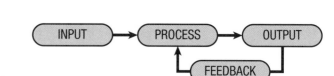

INPUT PROCESS OUTPUT

There are two basic types of control:

- an **open-loop** control – has no means of checking its output
- a **closed-loop** control – has a means of checking its output that is feedback.

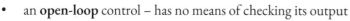

INPUT → PROCESS → OUTPUT

INPUT → PROCESS → OUTPUT → FEEDBACK

An open-loop control

A closed-loop control

If you look around your home, school or community, you should be able to find examples of both types of control system. Here is one example of a control system.

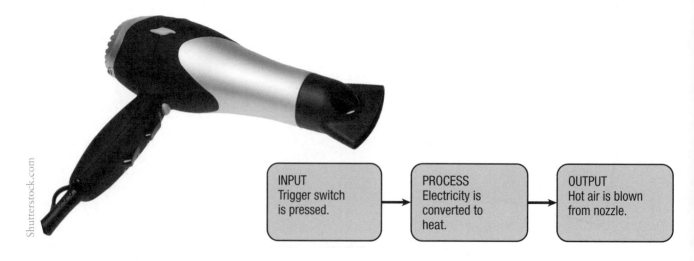

Shutterstock.com

INPUT
Trigger switch is pressed.

→

PROCESS
Electricity is converted to heat.

→

OUTPUT
Hot air is blown from nozzle.

Type of control: ....

Q1 Look at the items shown in the following photographs. Each of them works using a control system. Draw a control system diagram for each one and describe its elements (input, process and output – and feedback, where applicable). Then identify the type of control.

Door bell

Type of control: ....

Stereo system

Type of control: ....

Oven

Type of control: ....

Hot-water storage system

Type of control: ....

Q2 Today, many control systems are monitored by a computer. Through additional research and/or class discussion, identify two control systems that are monitored by computer. Then write a few sentences about each system and what it does.

1 ....

....

....

2 ....

....

....

Teacher's signature   **Date**

....

WORKSHEET

3.4 Mechanisms

These activities will help you to:

▶ gain knowledge of different types of mechanisms

▶ understand how these mechanisms work.

All mechanisms involve motion of some kind. They change one kind of motion into another kind of motion. There are four basic types of motion. Each type of motion has its own symbol, as shown.

Q1 Draw the symbol for each type of motion.

The symbol for **linear** motion is .... or ....

The symbol for **rotary** motion is ....

The symbol for **oscillating** motion is .... or ....

The symbol for **reciprocating** motion is .... or ....

MOTION SYMBOLS

Levers

A lever consists of a rigid bar that pivots on a fixed point called a fulcrum. There are three classes of lever: Class 1 levers, Class 2 levers and Class 3 levers.

Q2 Sketch diagrams in the boxes below to illustrate each class of lever. Identify the load (L), fulcrum (F) and effort (E). An example is given for each class of lever. Name at least two other devices that apply this principle.

Class 1 lever	Class 2 lever	Class 3 lever

....

....

....

....

Linkages

A linkage is formed by joining a number of levers. Linkages are used to transmit motion and force in a desired way.

Q3 Sketch a linkage diagram to show:

Input motion = output motion	**Large input motion, but small output motion**
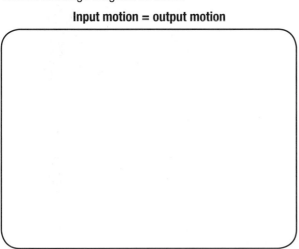	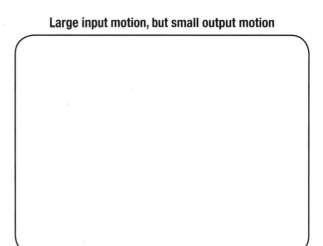

Identify the input motion, the pivot point and the output motion.

Q4 Through further research and/or class discussion, find three other types of linkage. List them here.

1 ....   2 ....   3 ....

Pulleys

Pulleys are another useful mechanism for transmitting force and movement. The basic form of a pulley is a grooved wheel. Pulleys are often linked by a belt.

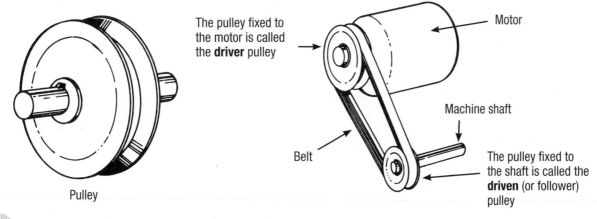

The pulley fixed to the motor is called the **driver** pulley

Motor

Machine shaft

Belt

The pulley fixed to the shaft is called the **driven** (or follower) pulley

Pulley

Q5 Describe one advantage and one disadvantage of using pulleys.

Advantage: ....

....

Disadvantage: ....

....

Q6 Name three machines in the home or workshop where pulleys are used to transmit force and movement.

1 ....

2 ....

3 ....

Pulley puzzles

Q7 What would happen if you used two pulleys with the same diameter as shown here? ....

....

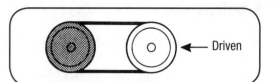
Driven

Q8 How could you use a belt to change the direction of rotation of the driven pulley? (You want it to rotate in the opposite direction to the driver pulley.) Sketch your answer on the diagram.

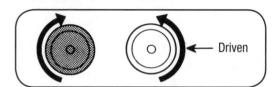
Driven

Q9 Match each numbered pulley system with the correct statement about its output. (In each illustration the driver pulley is shaded and on the left.)

- opposite direction, reduced speed ☐
- same direction, increased speed ☐
- opposite direction, increased speed ☐
- same direction, reduced speed ☐

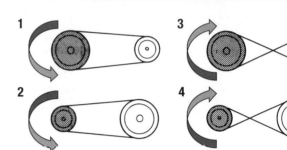

Gears

Gear wheels are designed to interlock smoothly or mesh with each other.

Gear wheel

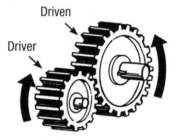
Teeth interlock (mesh)

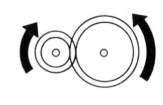
Driven
Driver

Q10 What is the main advantage of using gears instead of pulleys in a mechanism?

....

Gears meshed together form a **gear train**. There are two types of gear train:

- simple

- compound.

Q11 Draw sketches to show the two types of gear train. (Carry out research or discuss in class if necessary.)

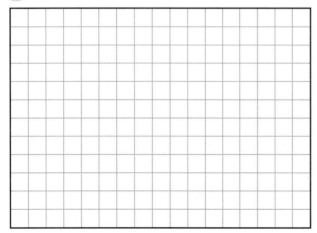
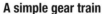
A simple gear train

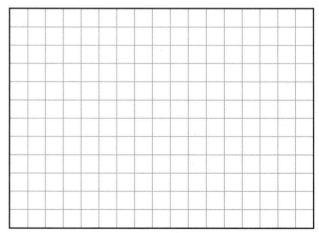
A compound gear train

To calculate the gear ratio of a simple gear train, you would use the formula:

$$\text{Gear ratio} = \frac{\text{Number of teeth on driven gear}}{\text{Number of teeth on driver gear}}$$

Q12 Circle the gear ratio that would be correct if a driver gear has 50 teeth and the driven gear has 125 teeth.

a 1 : 2.5   b 2.5 : 1   c 4 : 1   d 0.4 : 1

Cams

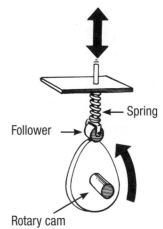

A rotary cam is a specially shaped wheel used to change rotary motion to linear motion. As the cam turns, it causes a follower to move in and out or up and down in a particular way. Differently shaped cams produce different patterns of movement in the follower.

Group activity

Q13 Each of the rotary cams shown below produces a different pattern of movement in the follower. Describe the movement of each follower as the cam turns one revolution. (You may wish to make a model to help you answer this question.)

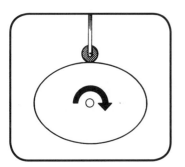

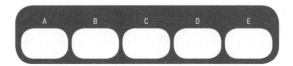

WORKSHEET

3.5 Electronics

These activities will help you to:

▶ gain knowledge of circuits and electronic components

▶ read a schematic circuit diagram.

Electronics involves connecting a number of components in an electrical circuit so that current can flow to produce a desired result.

Q1 These images show a number of electronic items. What desired result is produced by their electrical circuits?

Image 1   **Image 2**   **Image 3**

Image 1 ....

Image 2 ....

Image 3 ....

In electronics, a circuit consists of a continuous loop that can carry an electrical current.

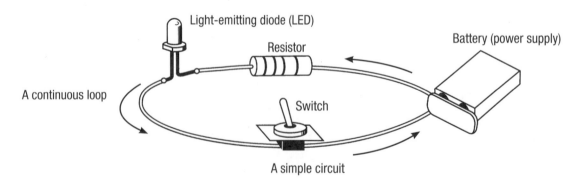

Q2 Which two conditions are necessary for electricity to flow in a circuit?

1 .... 2 ....

Electricity cannot flow through all materials. Materials that allow electricity to flow are called **conductors**. Materials that do not permit the flow of electricity are called **insulators**.

Q3 Write down two examples of a conductor.

1 .... 2 ....

Q4 Write down two examples of an insulator.

1 .... 2 ....

Q5 Through additional research (or class discussion), match the symbols and other abbreviations in Part A with their full values in Part B. (Use the letters from Part B to give your answers.)

Part A

emf (..........)
DC (..........)
V (..........)
pd (..........)
A (..........)
Ω (..........)
AC (..........)
M (..........)
m (..........)
I (..........)
R (..........)
k (..........)

Part B

a   potential difference

b   electrical current

c   prefix mega-, meaning one million (10^6)

d   alternating current

e   ohm (unit of resistance)

f   prefix kilo-, meaning one thousand (10^3)

g   electromotive force

h   ampere/amp (unit of current)

i   voltage

j   prefix milli-, meaning one thousandth (10^{-3})

k   direct current

l   resistance

Each component of a circuit has a specific function to perform. By joining components in a special way you can complete a circuit and make it work.

Every component has its own symbol. These symbols are related to what the component looks like and how they work.

A circuit design using these symbols is called a **schematic circuit diagram**.

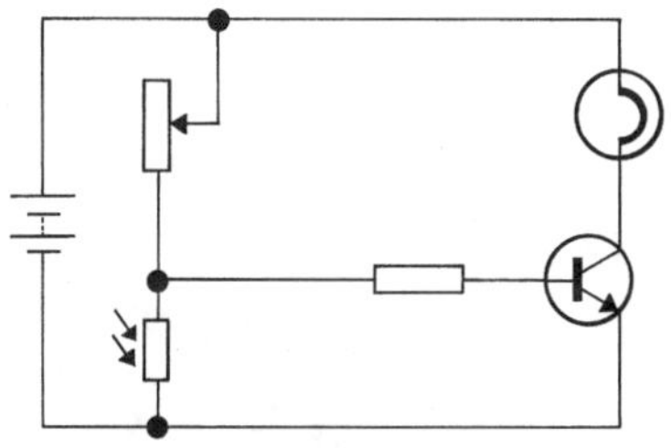

A schematic circuit diagram

Q6 Identify the components in the above schematic circuit diagram. (Refer to the symbol bank on p. 39.)

Q7 Select two components listed in the word bank (or use two of your own choice). Then write a sentence or two about the function of each component and also show its symbol.

Component 1 is .... Symbol:

Function: ....
....
....

Component 2 is .... Symbol:

Function: ....
....

WORD BANK
LED
N-P-N transistor
Polyester capacitor
Resistor
Rotary potentiometer
JFET transistor

Fixed-value resistors use coloured bands painted on the body to work out their resistance.

The coloured bands are part of a code in which each colour is assigned a value. Refer to the table on the next page.

The resistance of the 4-band resistor shown on the right is explained on the next page.

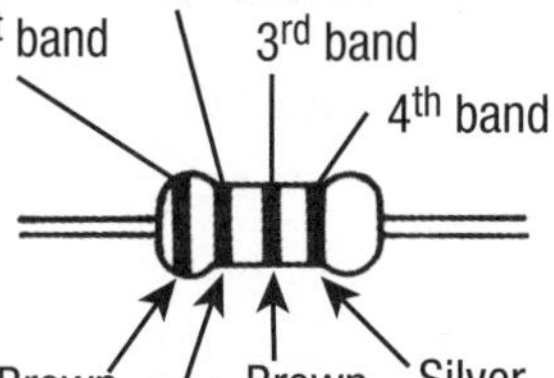

1st band
2nd band
3rd band
4th band

Brown Red Brown Silver tolerance band

Resistor colour code 4-band

Colour	1st band = 1st number	2nd band = 2nd number	3rd band = Zeros to add to the end	4th band = Tolerance band
Black	0	0	–	Brown = 1%
Brown	1	1	0	Red = 2%
Red	2	2	00	Gold = 5%
Orange	3	3	000	Silver = 10%
Yellow	4	4	0000	
Green	5	5	00000	
Blue	6	6	000000	
Violet	7	7	0000000	
Grey	8	8		
White	9	9		
To find the resistance of the example:	Brown = 1 ∴ 1st number is 1	Red = 2 ∴ 2nd number is 2	Add the zeros to the end: 0	This gives 120 ohms

The tolerance band is silver, giving a tolerance of ± 10%. This means that the resistor's value could be anywhere from 108Ω (120Ω – 10%) to 132Ω (120Ω + 10%).

Q8 What is the resistance of each of the resistors shown below? Also record their range.

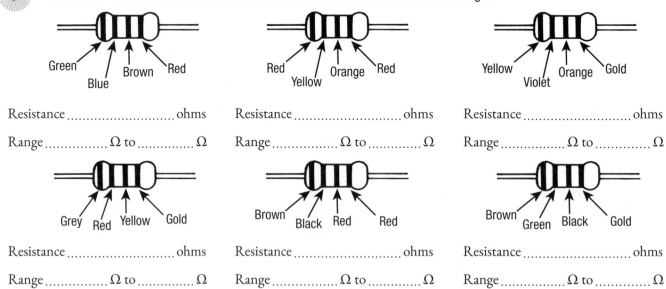

Green Blue Brown Red

Resistance .... ohms

Range .... Ω to .... Ω

Red Yellow Orange Red

Resistance .... ohms

Range .... Ω to .... Ω

Yellow Violet Orange Gold

Resistance .... ohms

Range .... Ω to .... Ω

Grey Red Yellow Gold

Resistance .... ohms

Range .... Ω to .... Ω

Brown Black Red Red

Resistance .... ohms

Range .... Ω to .... Ω

Brown Green Black Gold

Resistance .... ohms

Range .... Ω to .... Ω

Note: There are also 5-band and 6-band colour code systems used for higher tolerance resistors.

Components can be joined in a special way to complete a circuit and make it work (forming a **closed circuit**). They can be connected either **in series** (that is, one after another) or **in parallel** (that is, there are two or more paths that the current can take). If there is a break in the circuit, it is an **open circuit** and the current will not flow.

9780170439909 © 2019 Basil Slynko

Series circuits

In a series circuit, the flow of electricity (current) can only follow one pathway. The current must pass through all the components. Its value does not change.

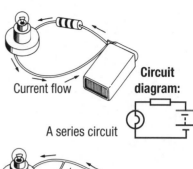

Current flow

Circuit diagram:

A series circuit

Parallel circuits

In a parallel circuit, there are two or more pathways that the current can take. The current is not the same throughout the circuit.

Q9 Label each of the six circuits below as a series and/or parallel circuit. Then state whether or not current is flowing in the circuit. If it is not, circle the problem and give a reason(s).

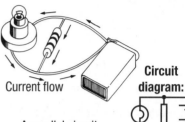

Current flow

Circuit diagram:

A parallel circuit

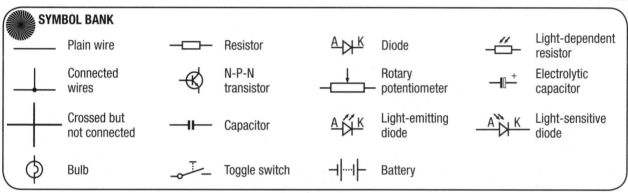

SYMBOL BANK

—— Plain wire	—▭— Resistor	A ▷⊦ K Diode	Light-dependent resistor
Connected wires	N-P-N transistor	Rotary potentiometer	Electrolytic capacitor
Crossed but not connected	—⊦⊦— Capacitor	A ▷⊦ K Light-emitting diode	A ▷⊦ K Light-sensitive diode
Bulb	Toggle switch	Battery	

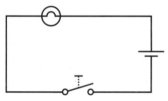

.... circuit

Will it work? ....

Why? ....

....

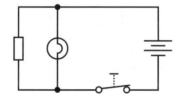

.... circuit

Will it work? ....

Why? ....

....

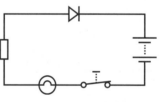

.... circuit

Will it work? ....

Why? ....

....

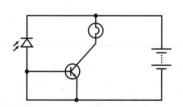

.... circuit

Will it work? _____

Why? _____

_____

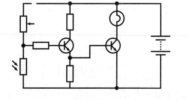

.... circuit

Will it work? _____

Why? _____

_____

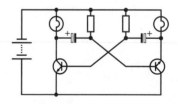

.... circuit

Will it work? _____

Why? _____

_____

WORKSHEET 9.6

Teacher's signature   Date

....

WORKSHEET

4.1 Materials

These activities will help you to:

▶ be aware of some of the different uses of materials through the ages

▶ gain knowledge of properties of materials

▶ learn about modern materials and their uses.

Q1 Through the ages, people have used many different materials for the purposes shown in these illustrations. Through research, identify the basic material(s) used in each of these applications.

....
....
....
....

....
....
....

....
....
....

....
....
....
....

....
....
....

....
....
....

....　..　..

....　..　..

....　..　..

Every material has a set of properties such as colour, strength and texture. These affect the appearance of the material (its aesthetic qualities) and also how it can be used.

 **STRENGTH:** How easy or difficult is it to break?	 **TOUGHNESS:** How much force can it stand without breaking?	 **HARDNESS:** How well does it resist denting or scratching?	 **ELASTICITY:** How well does it return to its original shape and size after being subjected to external loads and forces?
 **DENSITY:** How heavy or light is it for its size?	 **CONDUCTIVITY:** How well does it conduct (or insulate against) heat, electricity or sound?	 **DURABILITY:** How well does it resist environmental stresses such as the weather or insects, or does it tear easily or wear out (e.g. fabric)?	 **FLAMMABILITY:** How easily does it burn (or smoulder) when exposed to heat or fire?

The strength of a material can be measured in five ways.

Q2　List three of the ways below.

1..　2..　3..

Q3　What does it mean when a material has the property of:

• plasticity? Plasticity is

....

....

• biodegradability? Biodegradability is

....

....

In industry, the choice of material is very important. Manufacturers need materials that work properly, are safe to use, and are attractive to the purchaser. Often there are several materials that can be used for the same purpose.

A table

A skateboard

A candle stick

Q4 These pictures show three items. Select one and record the following information about the item.

- I have selected the ....

- Which materials have been used? ....

- What are some of the properties of these materials? ....

....

- Which other materials could have been used? Why? ....

....

A **composite** material is made by combining two or more materials such as glass fibre and epoxy resin. The aim is for the composite material to acquire properties that are superior to those of its components.

Composite materials can be grouped into three broad categories:

- layered composites (e.g. laminated windscreens and laminated fabric such as Gore-Tex®)
- particle composites (e.g. concrete)
- fibre composites (e.g. fibreglass and bonded fabrics).

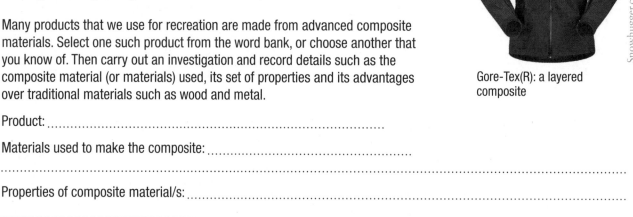

Gore-Tex(R): a layered composite

Q5 Many products that we use for recreation are made from advanced composite materials. Select one such product from the word bank, or choose another that you know of. Then carry out an investigation and record details such as the composite material (or materials) used, its set of properties and its advantages over traditional materials such as wood and metal.

Product: ....

Materials used to make the composite: ....

....

Properties of composite material/s: ....

....

Advantages over traditional materials: ....

....

 WORD BANK
Golf clubs
Running shoes
Snow skis
Tennis racquet
Windsurfer

 **WORKSHEET 9.7** ▶

Teacher's signature

Date

....

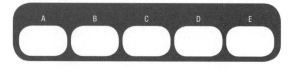

WORKSHEET

4.2

The anatomy of timber

These activities will help you to:

▶ know the steps in timber production

▶ know about tree growth and the different types of trees

▶ gain knowledge of the uses of forest products.

Timber production consists of six basic steps. They are:

1 **Forests**: softwood and hardwood forests are planted for commercial use. Some natural forests are also logged.

2 **Harvesting**: trees are felled with chain saws; branches are lopped and logs are trimmed.

3 **Raw materials**: logs are sorted into pulp logs, saw logs, ply logs, round timbers and forest thinnings.

4 **Processing**: the mills process the raw materials into different kinds of standard stock and treat against rot, etc. Waste is recycled.

5 **Products**: wood pulp, sawn timbers, veneers and plywood, round timbers, particle board and fibreboard are distributed to suppliers for sale.

6 **Timber users**: (for example, builders/carpenters, do-it-yourself (DIY) persons, furniture makers, wood turners, sculptors and landscapers) purchase various products for their own needs.

Q1 Use the information above to complete the following flow chart.

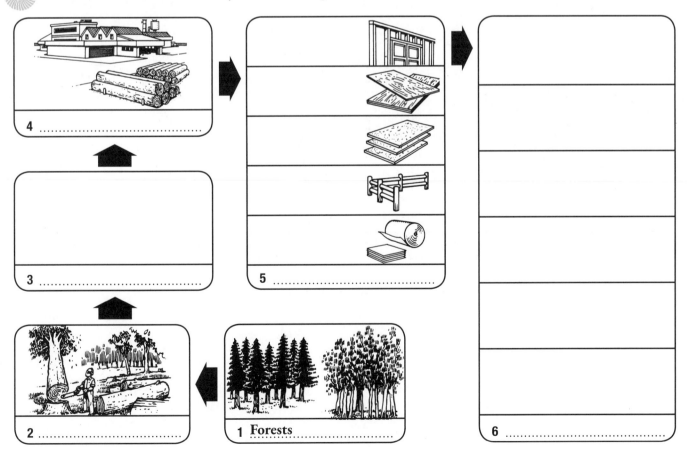

Q2 In the spaces provided, name the parts of the trees that are numbered in the diagrams below. (Carry out additional research if you need further information.)

1 ....

2 ....

3 ....

4 ....

5 .... wood

6 .... wood

7 ....

8 ....

9 .... ray

10 Hardwood/softwood* leaf

11 Hardwood/softwood* leaf

* Cross out the incorrect word.

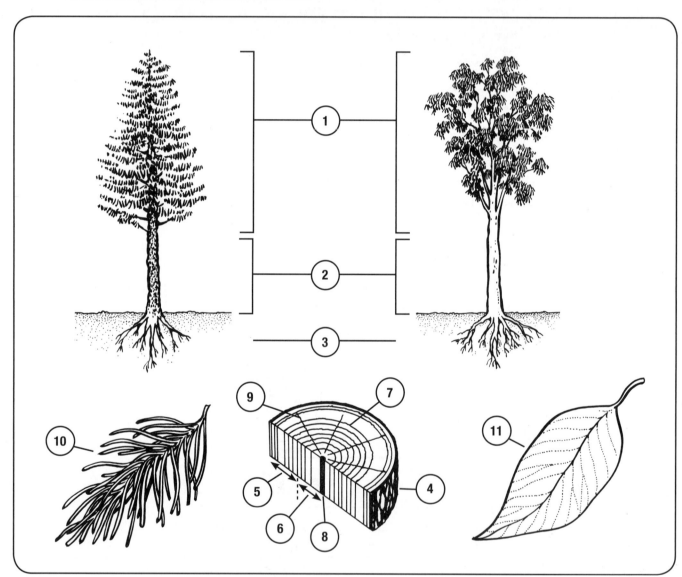

Add colour or rendering to the appropriate parts of the diagrams to make them look more realistic.

Q3 Why do we apply a 'finish' to timber products? Give at least two reasons.

....

....

....

....

....

9780170439909 © 2019 Basil Slynko

Forest products and by-products

Q4 The following table lists a number of timber products. For each product, fill in the missing details of uses, advantages and disadvantages.

To help you, details have been given for hardboard. Carry out research if you need further information.

Timber product	Use(s)	Advantage/disadvantage
Hardboard	Furniture construction	Weather-resistant. Can be curved. Hard on tools.
Particle board		
Plywood – interior, exterior – marine ply – structural		
Hardwood (e.g. gum)		
Softwood (e.g. pine)		
Laminated timbers		

Some forest by-products

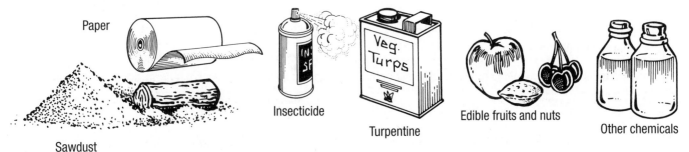

Paper

Sawdust

Insecticide

Veg. Turps

Turpentine

Edible fruits and nuts

Other chemicals

Q5 Choose three forest by-products from the above illustrations and enter information about them in this table. (An example is given.) Carry out research if necessary.

Forest by-product	Use(s)	Advantage/disadvantage
Methanol (wood alcohol)*	Fuel; chemical manufacture	Wide range of uses Highly flammable; poisonous

* Today, methanol is produced synthetically rather than from wood.

Teacher's signature   Date

....

A	B	C	D	E

WORKSHEET

4.3 Metals

These activities will help you to:

▶ know the properties of metals

▶ know the different uses for metals.

The properties of a metal determine how that metal can be used. If you make a poor choice of metal, the item you are making could break or malfunction in some way when it is used.

▶▶ INFORMATION FILE Properties of metals

The hardness of a metal is its ability to resist denting, scratching or wearing by abrasion.

When a metal is able to resist bending, twisting or impact without fracturing, it is said to have a high degree of toughness.

A metal is ductile when it can be permanently deformed by being stretched into wire.

The malleability determines how much it can be hammered or rolled without breaking.

A metal that has a high degree of conductivity can conduct heat and electrical current very well.

Annealing is the process of softening a metal by heating it so that internal stresses are reduced. Metal can be formed or machined more easily when it has been annealed.

Q1 Using the above information file, match each of the properties of metals in Part A with its correct definition in Part B. (Use the letters from Part B to give your answers.)

Part A

1 ductility   (.......)

2 hardness   (.......)

3 toughness   (.......)

4 malleability   (.......)

5 conductivity   (.......)

6 annealing   (.......)

Part B

a able to resist fracturing

b able to be hammered into sheets

c able to be drawn into wire

d the process of softening metal

e able to resist wearing

f able to conduct heat and electricity

▶▶ INFORMATION FILE Strength of metals

The strength of a metal can be measured in five ways. They are:

- **shear** strength: resistance to sliding forces acting in opposite directions

- **torsional** strength: resistance to twisting forces

- **compressive** strength: resistance to being squeezed together

- **bending** strength: resistance to a bending force

- **tensile** strength: resistance to being pulled apart.

Q2 Use the preceding information file to label the different kinds of strength that are being measured in these bars.

WORD BANK
Bending
Compressive
Shear
Tensile
Torsional

....

....

Metals are produced via manufacturing processes, using minerals and other raw materials. There are two groups of metals:

- **Ferrous** metals: those that are based on iron

- **Non-ferrous** metals: those that may contain, but are not based on iron.

Metals can be pure metals or alloys:

- A **pure metal** is one that has no other metal mixed with it.

- An **alloy** is a *mixture* of two or more metals. Alloys are often used when special properties are needed – for example, stainless steel (an alloy of steel, chromium and nickel) provides a high resistance to corrosion.

Q3 From your personal knowledge, or through further research and class discussion, fill in the blanks in the table below. (The first row has been completed as an example.)

Name of metal	Colour	F/NF*	PM/A*	Main property/ies	Common uses
Aluminium	Greyish white	NF	PM	Good thermal and electrical conductor; light but strong	Kitchenware, foil wiring; door/window frames; drink cans
Brass					
Copper					
Galvanised iron					
High-speed steel					
Lead					
Low-carbon steel					
Stainless steel					
Tin					

* F = ferrous, NF = non-ferrous, PM = pure metal, A = alloy

Teacher's signature   Date

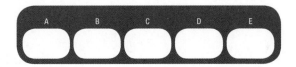

....

WORKSHEET

4.4

Know your plastics

This activity will help you to:

▶ know the difference between thermoplastics and thermosetting plastics

▶ know some of the different uses of plastics in everyday life.

Thermoplastics

Thermoplastics (also known as thermosoftening polymers) soften when they are moderately heated and harden when they cool. This process can be repeated many times. Every time the plastic is heated it will soften and return to its original shape. This is known as **plastic memory**. (Note that plastics can be damaged by overheating.)

Thermosetting plastics

Thermosetting plastics (also known as thermosetting polymers) are changed from one form to another by heat or chemical means. However, once the plastic sets hard it cannot be softened again (unlike thermoplastics).

▶▶ INFORMATION FILE

Some common plastics and their uses

Thermoplastics

Acrylic – advertising signs, car tail-lights

ABS – telephones, radio and TV cabinets, chairs, suitcases, consoles

Cellulose acetate – buttons, combs, ballpoint pens, photographic film, blister packaging

Nylon – fabrics, bearings, ropes, fishing line, zippers

Polycarbonate – street lighting covers, face shields, space helmets

Polyethylene – plastic film, pipe, bottle crates, lunch boxes

Polypropylene – spatulas, picnic ware, buckets, children's toys, woven sacks

Polystyrene – disposable cups, dairy food containers

Polystyrene foam – insulation, packaging for delicate equipment

Polytetrafluoroethylene (PTFE) – coating for non-stick utensils, plumber's pipe-jointing tape

Polyurethane foam – carpet underlay, upholstery, insulation (sound and refrigeration), head rests, surfboards

PVC (polyvinyl chloride) – pipe and conduit, coatings of fabrics and wire

Thermosetting plastics

Phenolic – bonding for plywood, particle board, abrasive wheels

Melamine formaldehyde – tableware, picnic ware

Urea formaldehyde – buttons, handles, knobs

Alkyds – paint, radio and TV components

Epoxy resins – adhesives, casting resin, moulds, surface coating, pattern making

Polyester resins – reinforced with glass fibre to make boats, surf skis, sheathing for surfboards

Q1 Use the information file to name the plastic used to make each of the items on the next page of this worksheet.

1 .... 2 .... 3 .... 4 ....
5 .... 6 .... 7 .... 8 ....
9 .... 10 .... 11 .... 12 ....
13 .... 14 .... 15 .... 16 ....
17 .... 18 .... 19 .... 20 ....
21 .... 22 .... 23 .... 24 ....
25 .... 26 Thermoplastic/Thermoset *   * Cross out the incorrect word

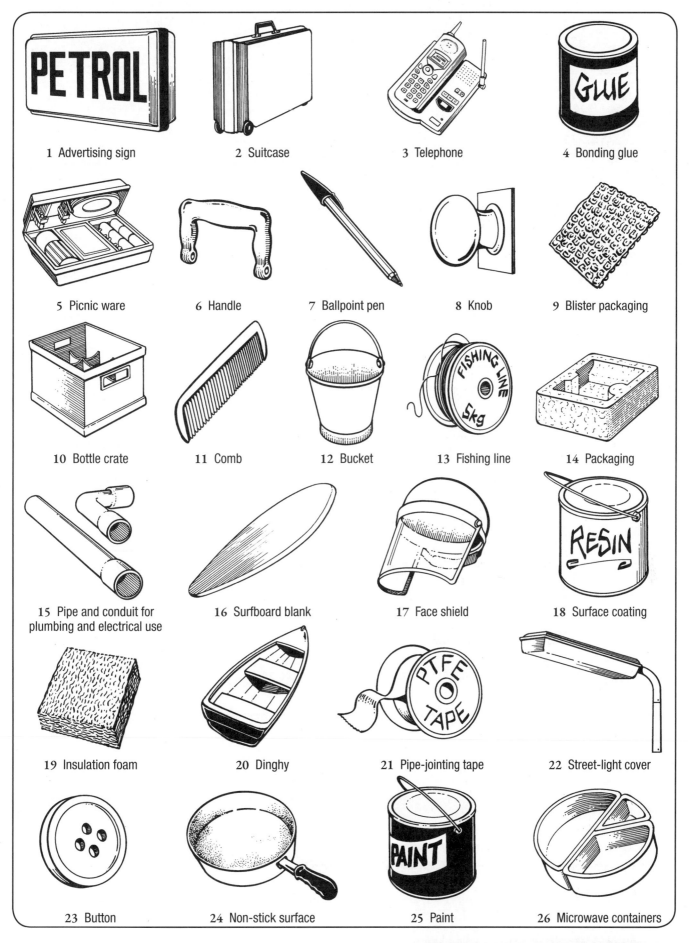

1 Advertising sign

2 Suitcase

3 Telephone

4 Bonding glue

5 Picnic ware

6 Handle

7 Ballpoint pen

8 Knob

9 Blister packaging

10 Bottle crate

11 Comb

12 Bucket

13 Fishing line

14 Packaging

15 Pipe and conduit for plumbing and electrical use

16 Surfboard blank

17 Face shield

18 Surface coating

19 Insulation foam

20 Dinghy

21 Pipe-jointing tape

22 Street-light cover

23 Button

24 Non-stick surface

25 Paint

26 Microwave containers

Teacher's signature   Date

....

WORKSHEET

4.5 Textiles

9780170439909 © 2019 Basil Slynko

These activities will help you to:

▶ gain knowledge of textiles

▶ know the properties of fibres.

A textile is a fabric or cloth that is made by weaving, knitting or bonding yarns together. The yarns can be made of natural and/or synthetic fibres that are spun to make the thread.

Q1 Use the word bank to identify the various parts of this fabric.

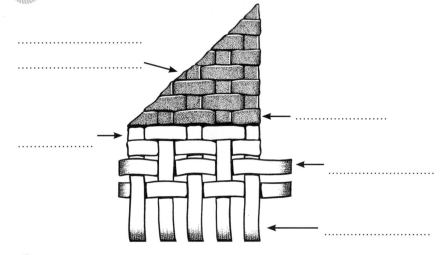

WORD BANK
Bias
Diagonal
Selvedge
Warp thread
Weft thread

Q2 Draw an arrow to indicate the straight grain of a fabric in the picture above.

Every fibre has its own set of properties. Fibres can be combined to create fabrics with specific properties. There are two ways to achieve this, either:

• as a **mixture**

or

• by **blending**.

Q3 Through research and/or class discussion, describe each term below.

A mixture is ....

....

Blending is ....

....

9780170439909 © 2019 Basil Slynko

There are advantages of combining different fibres into one fabric. A common combination is cotton and polyester, an example being 65% cotton and 35% polyester. The fabric then has the properties of both fibres.

Q4 Using the table 'Properties of fibres' on the next page, list some of the properties of a combined cotton/ polyester fabric.

....

....

....

....

Fabric samples

....

....

Q5 Through research and/or class discussion, list some other common combinations of fibres. (What are the advantages?)

....

....

....

When marking out fabrics, you need to line up the pattern with the warp grain to ensure it doesn't stretch out of shape (dimensional stability). On the other hand, if the fabric needs to be stretched, line up the pattern with the bias.

Pattern pieces should be arranged on the fabric in such a way that ensures there is minimum wastage.

Q6 Shown below are two patterns. With the aid of a sketch, show where and how you would place each pattern on the fabric to:

• ensure dimensional stability – Pattern A

• be stretched – Pattern B

• and ensure minimum wastage.

Selvedge

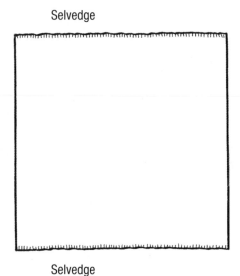

Selvedge

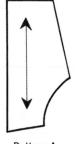

Pattern A
Number required: 3

Pattern B
Number required: 2

▶▶ INFORMATION FILE

Properties of fibres

Fibre	Main properties
Acrylic	good strength; very little stretch; resists creasing; low conductor (feels warm); fairly easy-care; melts
Cotton	durable; strong (wet or dry); water-absorbent; creases easily; good conductor (feels cool); very little elasticity; burns to ash
Linen	durable; very strong (wet or dry); water-absorbent; creases easily; good conductor (feels cool); no elasticity; burns to ash
Polyamide	very durable; very strong (wet or dry); does not crease easily; very poor conductor (feels warm); very high elasticity; easy care; burns and melts
Polyester	durable; water-repellent; good strength; resists creasing; fair conductor (feels cool); easy-care; burns and melts
Rayon	good strength; very absorbent; creases readily; good conductor (feels cool); burns and melts
Silk	durable; very strong; water-absorbent; resists creasing; very poor conductor (feels warm); reasonable elasticity; soft to touch; smoulders
Wool	moderately weak; water-absorbent; does not crease easily; very poor conductor (feels warm); very elastic; not insect-resistant; smoulders

Fragments of cloth have been found at a crime scene and some forensic tests have been carried out. The tests show that the fibres have the following properties: durability, strength and elasticity.

Q7 Using the above information file, can you identify the fibres? (There may be more than one possibility.)

My conclusion is ....

....

Q8 Which additional test(s) would you request to confirm the identity of the fibres?

....

Q9 What is your final conclusion on the identity of the fibres?

....

....

Smart materials react to external changes in their surroundings, such as light, temperature, electricity or pressure, and alter their state and return to their original state.

Q10 Through research and/or class discussion, investigate smart textiles. List some recent innovations below.

....

....

....

Teacher's signature   Date

....

WORKSHEET
4.6

Other common materials

This activity will help you to:

▶ become aware of the properties and uses of some different materials

▶ learn how to handle materials safely

▶ gain knowledge of Personal Protective Equipment (PPE) requirements.

Q1 Complete the table, on this page and next, using your personal knowledge or through research and class discussion.

The first row has been completed as an example.

Material	Some properties	Common use, with advantages and disadvantages	Any safety considerations?
Glass	brittle, hard, quite tough, non-conductor of electricity	**Common use:** windows, doors, drinking vessels, glass fibre products **Advantages:** transparent **Disadvantages:** poor resistance to impact	Yes ☐ No ☐ **Why?** risk of injury when handling **PPE requirements:** gloves, long sleeves, a thick cotton shirt and an apron
Rubber		**Common use:** .... **Advantages:** .... **Disadvantages:**	Yes ☐ No ☐ **Why?** .... **PPE requirements:** ....

Material	Some properties	Common use, with advantages and disadvantages	Any safety considerations?
Cork		**Common use:** .... **Advantages:** .... **Disadvantages:**	Yes ☐ No ☐ **Why?** .... **PPE requirements:** ....
Cement products		**Common use:** .... **Advantages:** .... **Disadvantages:**	Yes ☐ No ☐ **Why?** .... **PPE requirements:** ....
Ceramics/ baked clay		**Common use:** .... **Advantages:** .... **Disadvantages:**	Yes ☐ No ☐ **Why?** .... **PPE requirements:** ....

Teacher's signature   Date

....

WORKSHEET
5.1 Safety

These activities will help you to:

▶ become aware of the general need for safety

▶ be able to work out your own safety rules

▶ gain knowledge of the dangers of noise.

Safety is accident prevention. Accidents can happen anywhere, but they are common in the home, at work and on the sports field.

Select an activity that you enjoy.

My activity is ....

Q1 What are some of the silly or dangerous things that people could do when carrying out this activity? What injuries could result?

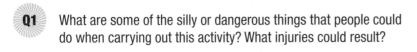

....

....

....

....

Q2 Write down four rules that you believe to be the most important for safety when carrying out this activity.

1 ....

2 ....

3 ....

4 ....

People are often injured as a result of carrying out **unsafe practices** or as a result of **unsafe conditions**.

• An unsafe practice is behaviour that could result in an accident.

• An unsafe condition is a situation in which an accident is likely to happen.

Unsafe practice

Unsafe conditions

Q3 Conduct a safety check at home, at school or in the community. List any unsafe practices or unsafe conditions that you can find. Then suggest a course of action for solving each problem. (Report your findings to someone in authority.)

Unsafe practice or condition	My solution
....	....
....	....
....	....
....	....
....	....
....	....
....	....

Personal safety relies on safe practices and safe conditions.

Q4 What do the letters PPE mean? (Hint: they stand for three words meaning safety gear.)

The letters PPE stand for ....

Q5 What do the letters SDS mean? (Hint: they stand for the name of a fact sheet about a hazardous substance.)

The letters SDS stand for ....

Q6 Describe the two areas that are addressed in an SDS.

1 ....

2 ....

Noise is a part of everyday life in industrial societies. The level of sound is measured in **decibels** (dB). You can permanently damage your hearing if you are exposed for too long to sound levels above 85 dB. However, you can protect your hearing by wearing earplugs and/or ear muffs where there is a high level of sound.

Q7 Rearrange the scrambled words in the following sentence so that they are correct and write them in the spaces provided.

You can avoid hearing damage if you wear ARE FUMSF (.................................) and/or

LAGUSREP (....) in areas where there is a high level of sound.

Above

80 dB
- Ear protection recommended

Above

85 dB
- Ear protection must be worn!

Ear Protection

Safety Signs & Product images are supplied by Seton Australia Pty Ltd

Q8 Indicate with a cross which of the following activities could lead to hearing damage.

Listening to a rock concert: 115 dB ☐

Talking: 65 dB ☐

Spray painting: 105 dB ☐

Using a power circular saw: 110 dB ☐

Standing near jet engines: 140 dB ☐

Listening to your personal stereo with headphones (e.g. MP3 player): 100 dB ☐

Using a jack hammer: 110 dB ☐

Using a food blender: 90 dB ☐

Name .... Date ....

Colours are used to attract attention. Colour is also used to convey safety messages. The four basic colours used in industry are red, yellow, green and blue. Each colour has its own safety message.

Q9 Look at the examples of safety signs. Use the word bank to identify each type of safety message.

Red

Yellow

Green

Blue

WORD BANK
Caution
Danger
Information
Safety

Safety message: ....

 Your State or Territory might have laws about how old you need to be before you can use certain machines and power tools. These age restrictions are for your safety. Always check with your teacher if you are unsure about what the law says.

Machines and power tools have moving parts that can cause serious injuries if you do not take care or if you use them incorrectly.

Q10 Complete the following 'true/false' safety quiz. (Circle 'T' or 'F'.)

You should always check a machine or power tool before you use it.　　T　　F

List at least three things that you need to check:

1 ....

2 ....

3 ....

All electrical machines and power tools must have a tag attached to the electrical cord, which must be up to date.　　T　　F

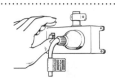

You should seek advice on the correct way to operate machinery or power tools.　　T　　F

Portable power tools can be carried by the electrical cord.　　T　　F

The cutting edge (e.g. a blade) must be clear of the material before you switch on the tool.　　T　　F

Power points can safely be piggy-backed as shown to the right.　　T　　F

 All materials need to be handled safely. This is for your own protection as well as that of other people.

Q11 Explain the safe practice for each case below.

Carrying long lengths of stock – light or heavy:

....

Carrying transparent sheet material such as glass:

....

Placing oversized materials on a work surface or bench:

....

WORKSHEET 9.8

Teacher's signature .... Date ....

Name.. Date...................................

5.2

Workplace health and safety

These activities will help you to:

▶ become aware of the general need for safety in the work environment

▶ know about safe working practices and safe working conditions.

There are many **hazards** (risks, dangers) hidden in this drawing.

Q1 Find at least 20 hazards. Number them on the drawing and describe them in the spaces provided opposite.

Hazards I found:

....
....
....
....
....
....
....
....
....

Q2 Now compare the hazards you have found with those found by some of your classmates.

Teacher's signature   Date

....

Workplace health and safety ▶ **55**

WORKSHEET
6.1 Tool families

These activities will help you to:

▶ gain knowledge of the functions and appropriate applications of different tools

▶ become aware of how tools have changed to meet current demands

▶ understand how tools are related to basic machines.

Today's tools belong to four main groups:

- **Hand tools**: any tool that requires only muscle power. Examples include hammers and scissors.

- **Portable power tools**: any easily transportable tool that requires an additional power supply such as electricity or compressed air. Examples include electric sanders, cordless drill/driver and pneumatic drills.

- **Universal equipment**: common lightweight equipment for general use in home workshops, school workshops and small businesses around the world. Examples include bench saws and disc belt sanding machines.

- **Industrial equipment**: powerful, heavy-duty equipment for special use in heavy manufacturing and construction industries. Examples include hydraulic presses, cranes and mining machines.

Q1 Give two reasons why industrial equipment needs to be sturdy or robust.

1 ....

2 ....

Q2 To which of the above four groups does each of these tools belong?

Oxyacetylene welding equipment

....

Disc sander

....

Ball pein hammer

....

Crane

....

Spring clamps

....

Bench drill

....

Portable jig saw

....

Cordless drill

....

9780170439909 © 2019 Basil Slynko

All tools can be classified according to the particular function they perform. The five basic **families** of tools are:

1 **marking-out/measuring** tools such as pencils and steel rules

2 **cutting** tools such as saws and trimming scissors

3 **clamping and holding** devices such as clamps and locking pliers

4 **percussion** (impact) tools such as hammers and mallets

5 **torsion** (twisting) tools such as spanners and screwdrivers.

Q3 Prepare a mind map for hand tools. List some examples other than those already given.

Early human beings developed tools from stone, wood and bone to help them perform daily tasks. These tools were modified and adapted over time as a result of changes in materials and energy sources.

From prehistoric implements to modern power tools

Q4 Select one such tool and carry out research to trace the sequence of development from its earliest form to the present day. (You may wish to draw a timeline or present a mind map.)

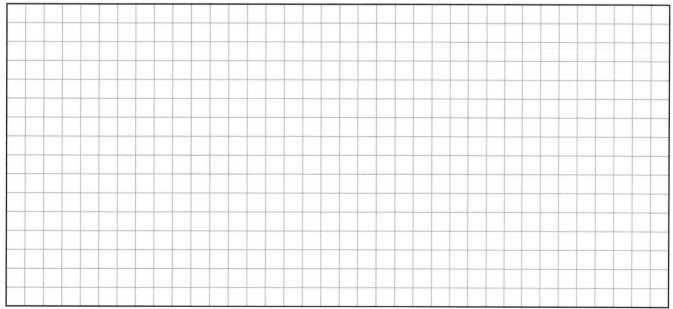

There are six basic machines: the lever, the pulley, the wheel, the screw, the wedge and the inclined plane. They multiply the force applied or change the speed produced by a person. All tools are related to these basic machines.

Q5 Use the word bank to identify the basic machine(s) to which each of the following tools is related.

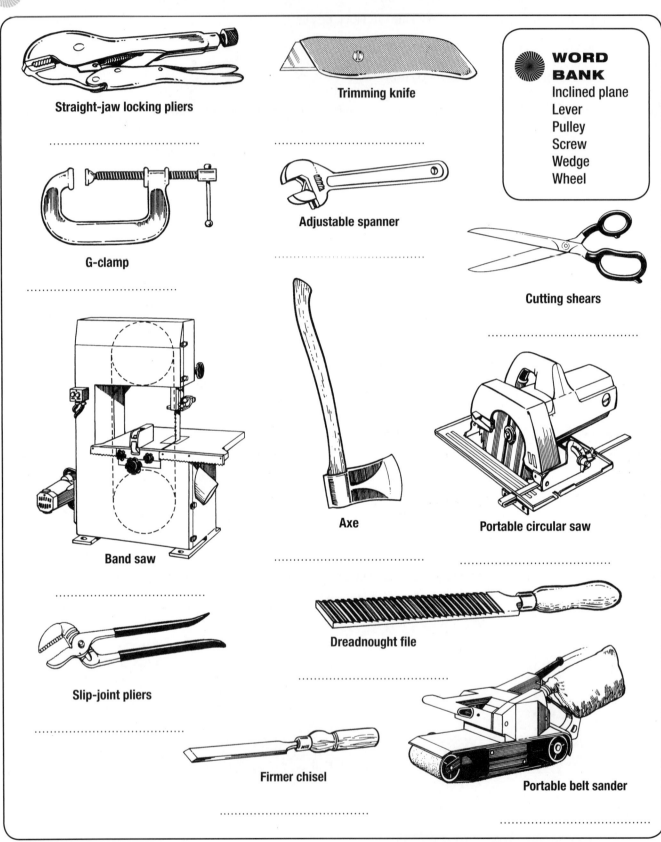

Straight-jaw locking pliers

Trimming knife

WORD BANK
Inclined plane
Lever
Pulley
Screw
Wedge
Wheel

G-clamp

Adjustable spanner

Cutting shears

Band saw

Axe

Portable circular saw

Slip-joint pliers

Dreadnought file

Firmer chisel

Portable belt sander

Teacher's signature   Date

WORKSHEET

6.2

Tools – working with timber

These activities will help you to:

▶ know the names and uses of tools

▶ know some different families of tools

▶ be able to put different tools into their correct families.

Q1 Look at the tools shown in this drawing. Place numbers from the word bank in the circles provided to identify the tools.

WORD BANK

1 Bench clamp
2 Bench hook
3 Cabinet screwdriver
4 Chisels
5 Claw hammer
6 Coping saw
7 Folding rule
8 G-clamp
9 Hand drill
10 Mallet
11 Marking gauge
12 Mortice gauge
13 Nail punch
14 Pencil
15 Phillips head screwdriver
16 Pincers
17 Plane
18 Quick-grip clamps
19 Rasp
20 Rose countersink
21 Sash cramp
22 Sliding bevel
23 Spade bit
24 Spring clamp
25 Surform
26 Tape measure
27 Tenon saw
28 Try square
29 Twist drill
30 Warrington hammer
31 Webbing clamp
32 Woodworker's vice
33 Workmate®

Some of the tools shown on the previous page can also be used when working with other materials, such as plastics.

Q2 Identify these tools by placing the capital letter 'P' next to them.

Q3 Now put the names of some tools from the word bank in the correct column of this table of tool families. (Five examples have been given.)

Tool families				
Marking-out/ measuring tools	Cutting tools	Clamping and holding tools	Percussion (impact) tools	Torsion (twisting tools)
Folding rule	*Tenon saw*	*G-clamp*	*Mallet*	*Screwdriver*

Q4 Woodworking tools crossword

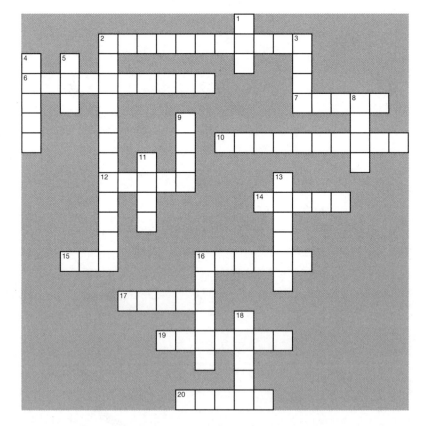

Across

2 Used with screws
6 This hammer is used with brads
7 Puts heads of nails below the surface
10 Used to finish curved surfaces
12 The partner of a bit
14 Used to mark the distance from an edge
15 Makes a hole for screws or nails
16 Used to remove waste to make grooves
17 Holds work when glued
19 Used to remove brads or nails
20 Used to reduce timber

Down

1 Used with a brace to bore holes
2 A ... ... is used to mark angles on timber
3 A file that removes wood
4 ... drills are used for boring holes
5 A ... square is used to draw a line at right angles to an edge
8 A hammer used with nails
9 Used for measuring
11 A ... drill
13 Used with wood chisels when removing waste
16 A ... saw is used for cutting curved shapes
18 A saw commonly used in woodworking

Teacher's signature   Date

....

WORKSHEET
6.3

Tools – working with metals

These activities will help you to:

▶ know the names and uses of tools

▶ know some different families of tools

▶ be able to put different tools into their correct families.

Q1 Look at the tools in this drawing. Place numbers from the word bank in the circles provided to identify them.

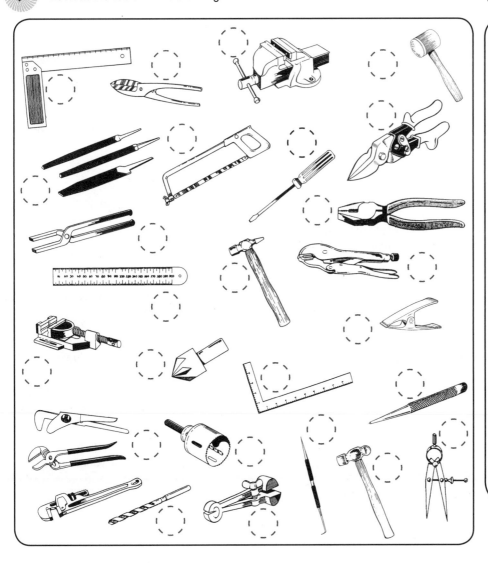

WORD BANK
1 Aviation snips
2 Ball-pein hammer
3 Centre punch
4 Combination pliers
5 Cross-pein hammer
6 Dividers
7 Engineer's screwdriver
8 Engineer's square
9 Engineer's vice
10 Files
11 Hacksaw
12 Hand vice
13 Hole saw
14 Locking pliers
15 Machine vice
16 Rose countersink
17 Scriber
18 Soft-faced hammer
19 Spring clamp
20 Steel rule
21 Tin snips
22 Tinman's square
23 Tongs
24 Twist drill
25 Wrenches

Some of the tools shown above can also be used when working with other materials, such as plastics.

Q2 Identify these tools by placing the capital letter 'P' next to them.

Q3 Now put the name of some tools from the word bank in the correct column of the table of tool families below. (Four examples have been given.)

Tool families				
Marking-out/ measuring tools	Cutting tools	Clamping and holding tools	Percussion (impact) tools	Torsion (twisting) tools
Steel rule	Hacksaw	Engineer's vice	Soft-faced hammer	

When we make an article, we use one of the following three processes to transform the material being used:

- **Forming**: the process of changing the shape of the material without removing any part of it. Examples of forming processes include bending, stretching and twisting metal.

Q4 Name one tool used for **forming** metal: ....

- **Separating:** the process of removing part of the material from the parent material. Examples of separating processes include cutting, filing, drilling and shearing metal.

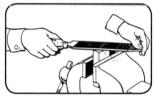

Q5 Name one tool used for **separating** metal: ....

- **Combining:** the process of joining two or more similar or dissimilar materials. Examples of combining processes include soldering and riveting metals.

Q6 Name one tool used for **combining** metal: ....

Q7 All of the items below are used in forming sheet metals. Identify them by placing numbers from the word bank in the circles provided.

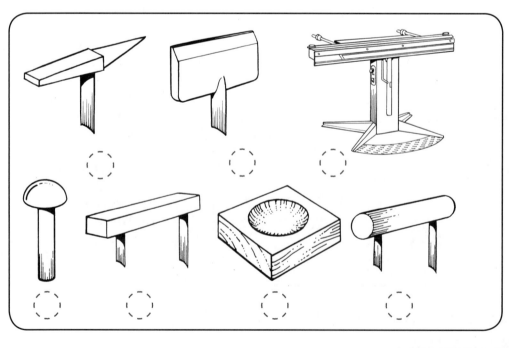

WORD BANK
1 Bick/beck/beak iron
2 Hatchet stake
3 Hollowing block
4 Mushroom stake
5 Round mandrel
6 Sheet metal folder/ Pan brake/Press brake
7 Square stake

Teacher's signature   Date

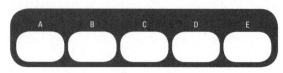

....

WORKSHEET
6.4

Portable power tools and machinery

These activities will help you to:

▶ know how to handle power tools and machinery safely

▶ know some of the features of power tools and machinery.

 Some portable power tools may have an age restriction in your State or Territory. Legislation may prevent you from using this tool. This is for your safety! Check with your teacher. (Note: Any restricted portable power tools described below are presented as general information.)

Power tools and machinery can be used to increase the output of work in a given period; they can also improve the quality of work. They are used in the manufacturing and construction industries, and in home workshops.

 However, the cutting implement (for example, the saw blade or the router bit) cannot tell whether it is cutting timber, metal, plastic, flesh or bone, so great **CAUTION** must be exercised at all times when using power tools and machinery.

Q1 List two advantages of using power tools and machinery.

1 .... 2 ....

Portable electric drill

Q2 Use the word bank to name the parts indicated.

 WORD BANK
Chuck/Keyless chuck
Cooling vents
Cord strain reliever
Pistol-grip handle
Reversing switch
Trigger switch/variable
speed switch

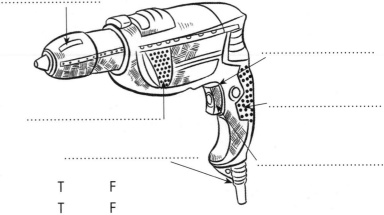

Q3 True or False?

Work should be firmly secured when drilling.   T   F

Press as hard as you can when using a drill.   T   F

Q4 When purchasing a portable electric drill, you need to consider:

1 chuck size

2 speed control

3 other? ....

Q5 Through research and/or class discussion, explain briefly why each of the above factors needs to be considered.

1 ....

2 ....

3 ....

Cordless jig saw

Q6 Use the word bank to name the parts indicated.

WORD BANK
Battery
Blade
Blade-release lever
Handle
Pendulum switch
Table
Trigger switch
Variable-speed switch

Q7 True or False?

Selection of blade type is not important.	T	F
It is possible to cut any thickness of material.	T	F
Allow the blade to stop before removing it from the material.	T	F
The table must be in contact with the material at all times.	T	F

Q8 Through research and/or class discussion, determine which specific features you need to consider when purchasing a cordless jig saw. List them here.

....

....

Q9 Use the word bank to identify the jig-saw blade in the table below that you would use for cutting plastic, wood or metal.

WORD BANK
Metal
Plastic
Wood

Jig-saw blade	Material

Portable random orbital sander

Q10 Use the word bank to name the parts indicated.

WORD BANK
Dust box
Front grip
Handle
Pad
Trigger switch
Variable-speed dial

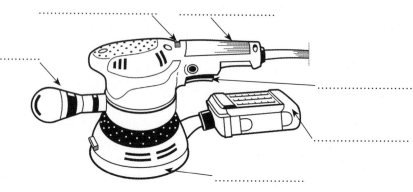

Q11 True or False?

Check the dust box before switching on   T   F

Before switching on the power tool, hold the tool firmly on the material surface.   T   F

Remove the sander from the work before switching it off.   T   F

Q12 When purchasing a portable random orbital sander, you need to consider:

1 a suitable pad area

2 the type of orbital action required

3 other? ....

Q13 Through research and/or class discussion, explain briefly why each of the preceding factors needs to be considered.

1 ....

2 ....

3 ....

Portable electric belt sander

Q14 Use the word bank to name the parts indicated.

WORD BANK
Belt
Belt tracking adjustment
Dust bag
Front grip
Handle
Trigger switch

Q15 True or False?

The type of belt used is an important factor in safe operation.   T   F

The abrasive belt can be fitted to the rollers in either direction.   T   F

Q16 Through research and/or class discussion, determine which specific features you need to consider when purchasing a portable electric belt sander. List them here.

....

....

Portable electric angle grinder

Q17 Use the word bank to name the parts indicated.

WORD BANK
Abrasive disc
Cooling vents
Guard
Side handle
Switch

Q18 True or False?

The abrasive disc must be securely clamped   T   F

The guard may be removed to fit larger diameter abrasive discs   T   F

Work should be firmly secured when grinding   T   F

Q19 Through research and/or class discussion, determine which specific features you need to consider when purchasing a portable electric angle grinder. List them here.

....

....

Electric drill press – bench or pedestal

Q20 Use the word bank to name the parts indicated.

WORD BANK

Belt guard	Depth stop	Motor
Chuck	Feed handle	Table
Column	Lower table/base	Table locking clamp

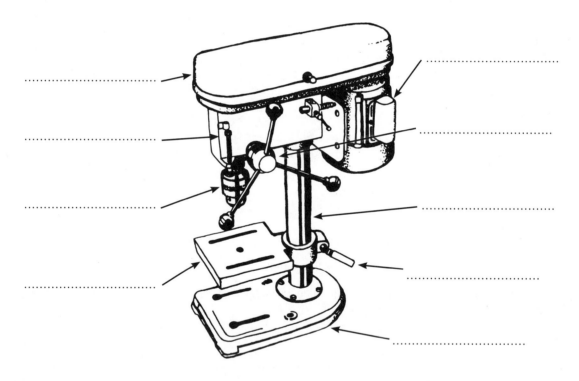

Q21 True or False?

Leave the chuck key in the chuck while the machine is operating.   T   F

The work does not have to be secured.   T   F

Different sizes of drill require different speeds.   T   F

The depth stop is used to indicate when you have drilled into the table.   T   F

Q22 Through research and/or class discussion, determine which specific features you need to consider when selecting an electric drill press. List them here.

....

....

Electric disc finishing sander

Q23 Use the word bank to name the parts indicated.

WORD BANK

Abrasive disc	Stand
Dust-collection hood	Switch
Motor	Table

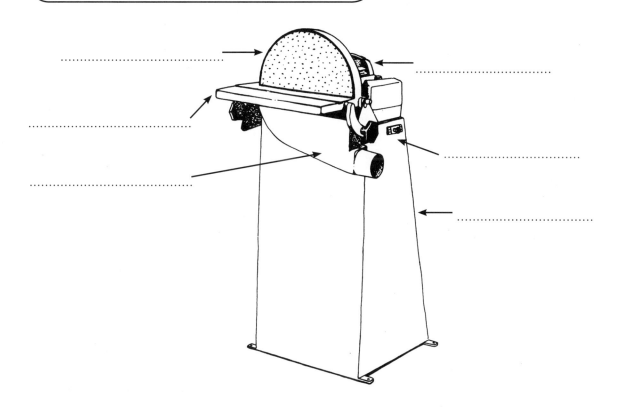

....

....

....

....

....

Q24 True or False?

ROTATION

Sanding should be carried out on both the left and the right side of the rotating disc.　　　　T　　F

Adjust the table to within 1 millimetre of the disc.　　　　T　　F

To stop the disc, you should switch off the power and slow the disc down by holding a piece of scrap timber against it.　　　　T　　F

Q25 When purchasing an electric disc sander you need to consider:

1 disc size

2 additional attachments

3 other? ....

Q26 Through research and/or class discussion, explain briefly why each of the above factors needs to be considered.

1 ....

2 ....

3 ....

Cordless power tools

Cordless power tools have become a popular alternative to portable power tools. They are quick and easy to use, and safer (no leads).

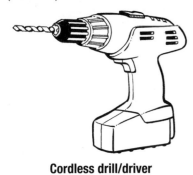

Cordless drill/driver

Cordless jigsaw

Cordless saw

When purchasing a cordless power tool, you need to consider:

1 application

2 size of battery

3 rate of recharge

4 cost and quality.

Note: There is another cordless system. It uses a gas. The gas is stored in disposable fuel cells.

Q27 Through research and/or class discussion, explain briefly why each of the factors listed above needs to be considered.

1 ....

2 ....

....

3 ....

....

4 ....

....

The range of cordless power tools is expanding to meet new applications.

Q28 Through research and/or class discussion, investigate the range of cordless power tools available in stores. List some of the range below.

....

....

....

....

....

....

WORKSHEET 9.9

Teacher's signature   Date

....

9780170439909 © 2019 Basil Slynko

Name.. Date..

WORKSHEET

7.1 Processes

These activities will help you to:

▶ be aware of a range of forming processes

▶ be able to select appropriate processes for separating materials

▶ be able to select appropriate processes for joining materials.

Materials can be transformed by three basic processes:

- forming – sometimes known as deforming

- separating – sometimes referred to as wasting or subtractive manufacturing

- combining – sometimes referred to as fabricating, assembling or additive manufacturing.

Forming is the process of changing the shape of material without removing any part of it.

Q1 Use the word bank to identify these forming processes. Write your answers in the spaces provided.

 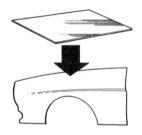 

> **WORD BANK**
> Casting/moulding
> Extruding
> Felting
> Forging
> Pressing
> Rolling
> Spinning

....

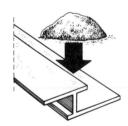 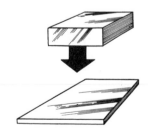  

....

Q2 The following illustrations show two articles and two heating units. Match each article with the heating unit used to form it.

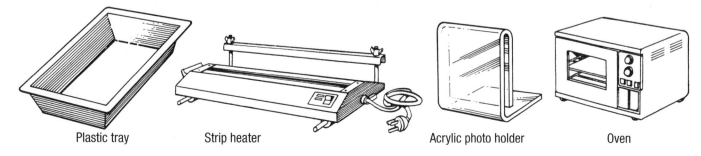

Plastic tray   Strip heater   Acrylic photo holder   Oven

Separating is the process of removing part of a material from the parent material. Examples include machining, shearing and abrading.

Q3 Which separating process listed in the word bank would you use for each of these applications? Record your answers in the spaces provided. State also the tool(s) that you would use.

To remove unwanted stock

To produce a flat surface on the end of a hexagonal rod

WORD BANK
Abrading
Cutting
Drilling
Filing
Planing
Polishing
Sawing
Turning

....
....
....

....
....
....

To remove small scratches or irregularities from a surface

To make a hole

To shape and finish material to an outline

....
....
....

....
....
....

....
....
....

To produce a copy of a pattern

To produce a conical shape

To reduce the thickness of stock

....
....
....

....
....
....

....
....
....

Approximate drill speed in rpm* for high-speed steel drill bits						
Material to be drilled	Drill diameter (mm)					
	1.5	3	6	8	10	12
Mild steel	5200	2600	1300	975	775	650
High-carbon steel	2600	1300	650	485	385	325
Aluminium alloys	6500	3250	1625	1200	975	800
Brass	6500	3250	1625	1200	975	800
Grey cast iron	5200	2600	1300	975	775	650
Acrylic	7000	3720	1800	1400	1110	900

* rpm = revolutions per minute

When drilling, thought must be given to the speed at which a drill should be operated. Refer to the table above for the next two questions.

Q4 What is the appropriate drill speed for each high-speed steel bit when:

- drilling a 3 mm hole in a piece of mild steel: .... rpm
- drilling a 3 mm hole in a piece of acrylic: .... rpm
- drilling a 3 mm hole in a piece of aluminium alloy: .... rpm
- drilling a 6 mm hole in a piece of mild steel: .... rpm
- drilling a 6 mm hole in a piece of brass: .... rpm
- drilling a 6 mm hole in a piece of acrylic: .... rpm
- drilling an 8 mm hole in a piece of high-carbon steel: .... rpm
- drilling an 8 mm hole in a piece of brass: .... rpm
- drilling an 8 mm hole in a piece of grey cast iron: .... rpm
- drilling a 10 mm hole in a piece of aluminium alloy: .... rpm
- drilling a 10 mm hole in a piece of mild steel: .... rpm
- drilling a 10 mm hole in a piece of brass: .... rpm
- drilling a 12 mm hole in a piece of high-carbon steel: .... rpm
- drilling a 12 mm hole in a piece of acrylic: .... rpm
- drilling a 12 mm hole in a piece of mild steel: .... rpm

Q5 Complete the following general rules about the speed of a drill. Cross out the incorrect word.

- Rule one: the larger the diameter of the drill, the (faster/slower) the drill speed.
- Rule two: the harder the material, the (faster/slower) the drill speed.

Combining is the process of joining two or more similar or dissimilar materials. Examples of combining methods include:

- mechanical fasteners – using screws, bolts, press-studs, and so on
- chemical adhesion – using glues and adhesives or solder
- jointing – through interlocking of the components
- cohesion – by means of solvents or welding.

Q6 Through further research and/or class discussion, record the different types of combining techniques (for example, nut and bolt, reclosable fasteners, through housing or contact adhesive) that you could use to combine the materials in these situations.

Where you give a number of possible techniques, asterisk (*) the one that you think is best.

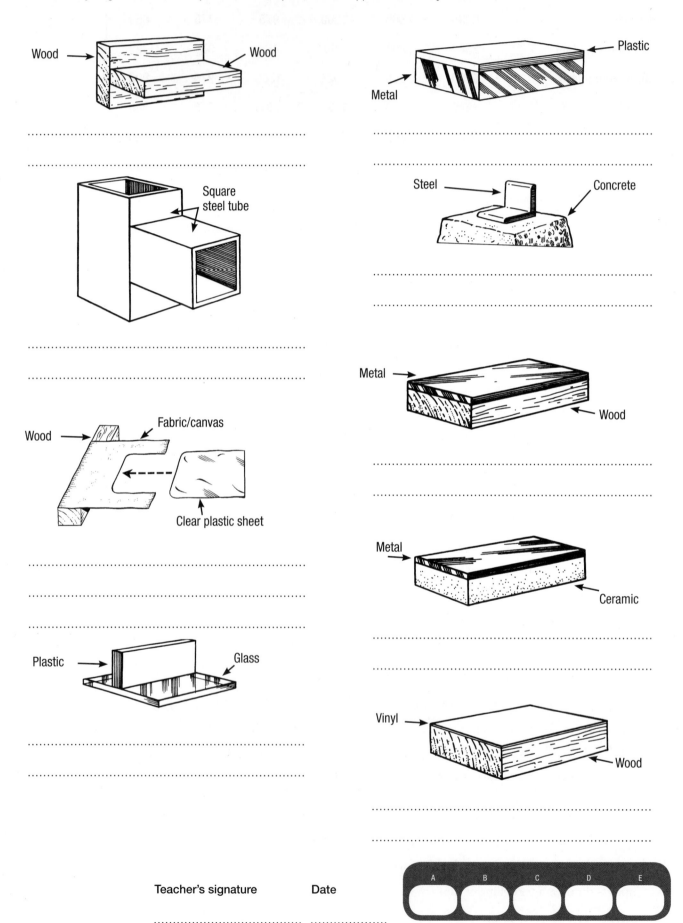

....

....

....

....

....

....

....

....

....

....

....

....

....

....

....

....

....

....

....

....

....

....

Teacher's signature   Date

....

| A | B | C | D | E |

9780170439909 © 2019 Basil Slynko

WORSHEET 7.2

Fasteners for timber

These activities will help you to:

▶ be able to identify different kinds of wood screw

▶ gain knowledge of the different uses for wood screws

▶ be able to classify wood screws into functional groups (according to their uses).

Q1 From the word bank, select the correct name of each type of wood screw shown below.

....

....

....

....

....

WORD BANK
Coach
Countersunk head
Particle board
Raised head
Round head

Q2 Match each type of screw in Part A with its correct use(s) in Part B. (Use the letters from Part B to give your answers.)

Part A: Type of screw

1 **Round head**   (.......)

2 **Countersunk head** (.......)

3 **Raised head**   (.......)

4 **Coach**   (.......)

5 **Particle board**   (.......)

Part B: Use(s)

a **For heavy work such as gate hinges and fastening down light machines**

b **To screw into particle board**

c **To fix metal to wood where the metal is too thin to be countersunk**

d **For decorative work, such as handles and fittings on furniture**

e **General purpose – head of screw to be flush with, or below, the surface**

Q3 List three metals from which screws are made.

1 ....   2 ....   3 ....

Q4 Use the word bank to name the numbered parts of the screws in this diagram.

1 ....

2 ....

3 ....

4 ....

5 ....

WORD BANK
Core or root diameter
Head
Length
Shank
Thread

This is an order for wood screws:

200 6G × 20 mm (6G × 3/4)* countersunk particle board wood screws, zinc-plated steel

The letter G stands for **gauge**. The gauge is a measure of the diameter of the screw shank. The higher the gauge number, the larger the diameter.

*Discuss in class why the screw size is given here in two forms.

Q5 When ordering wood screws, there are five important factors to keep in mind. What are they?

1 ....

2 ....

3 ....

4 ....

5 ....

When drilling holes for a screw to join two pieces of timber, you should follow the guidelines given in this diagram. As with nails, the thin piece of material is screwed to the thicker piece.

To select the correct length of screw, use this formula:

Length of screw = Thickness of Piece A × 3

Countersink if needed

Drill this clearance hole slightly bigger than screw shank

Drill this pilot hole about same size as core or root diameter.
The depth of the clearance hole plus pilot hole should equal the length of the screw

PIECE A

PIECE B

Q6 Circle the screw length you should use to join Piece A to Piece B. The thickness of Piece A is 11 millimetres. The screw is 10G and it is sold in these lengths:

a 25 mm

b 30 mm

c 40 mm

Q7 If this screw is going to be used to join two pieces of timber, what are the diameters of the holes you would drill?

Size of clearance hole for shank: .... mm

Size of pilot hole for core or root: .... mm

9780170439909 © 2019 Basil Slynko

Q8 Write one or two sentences to explain why the shank hole must be wider than the core or root hole.

....

....

Today, screws are available in a range of different types of heads.

Q9 Use the word bank to match each driver bit and head of a screw.

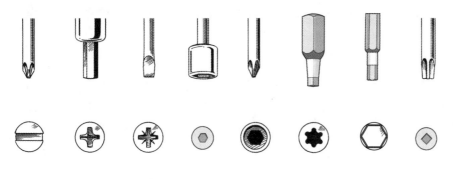

WORD BANK
- Hexagon
- Phillips
- Pozidriv
- Socket
- Square
- Straight slot
- Torx

Q10 Draw a line from each type of screw to the application that suits it best.

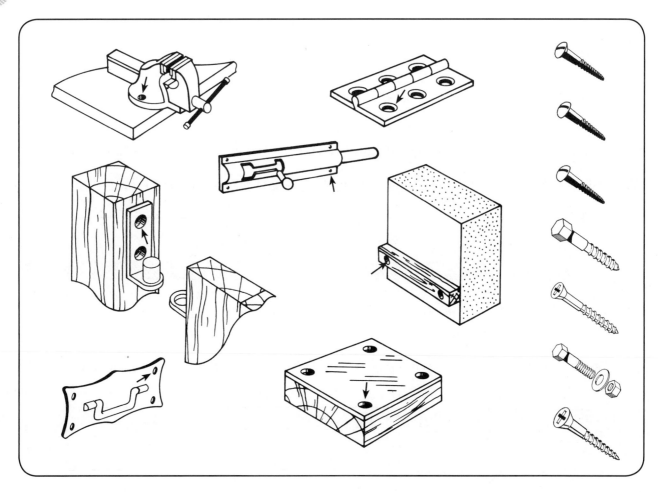

WORKSHEET 9.2

Teacher's signature

Date

....

9780170439909 © 2019 Basil Slynko

Fasteners for timber ▶ 75

WORSHEET
7.3 Fasteners – nails

These activities will help you to:

▶ gain knowledge of methods of nailing

▶ know which kind of nail to use for a particular purpose.

This is an order for nails: (1 kg   30 × 1.6 bullet-head steel wire nails)

Q1 When ordering nails, there are five important factors to keep in mind. What are they?

1 ....   4 ....

2 ....   5 ....

3 ....

Q2 List the metals from which nails are made.

....

When nailing two pieces of timber of unequal thickness, it is better to nail the thin piece to the thicker piece.

Q3 Is diagram A or diagram B the correct way to join the two pieces of timber? Circle your answer.

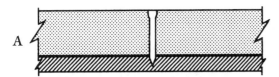 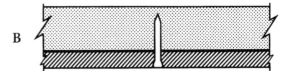

Timber has a grain. If nails are placed too close to each other along the grain, the timber may split. Therefore you have to be very careful when nailing near the end of a piece of timber.

Q4 Sketch in the grain on the pieces of timber in diagrams C, D and E. Which of the three diagrams shows the correct method of nailing? Circle your answer.

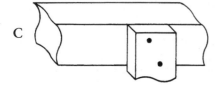 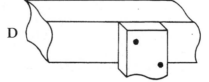 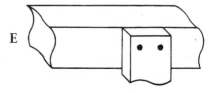

Q5 There are other methods of avoiding splitting when nailing close to the end of a piece of timber. Research two methods and describe them.

....

....

....

Q6 Use the word bank to name the numbered parts of the nail in this diagram.

1 ....

2 ....

3 ....

4 ....

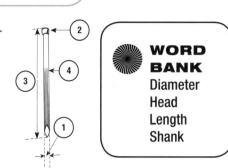

WORD BANK
Diameter
Head
Length
Shank

Nails are made in a variety of lengths. To select the correct length of nail, you need to know the thickness of the material to be nailed into place.

The general rule is to select a nail length at least **three times** this thickness. On no account should the nail length be less than twice this thickness. Always nail the thin piece of material to the thicker piece.

Q7 In the diagram below, circle the size of nail that should be used to fix piece A to piece B.

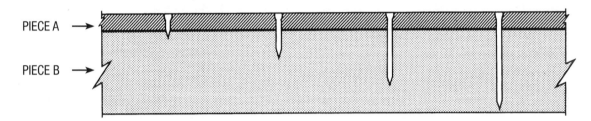

PIECE A ➞

PIECE B ➞

By driving nails perpendicularly, you rely on the grain fibres to hold the nail in the timber.

By sloping the nails in opposite directions, two pieces of timber can be held together much more strongly. This is called 'dovetail' nailing.

Q8 On this diagram, draw nails to show what is meant by 'dovetail' nailing.

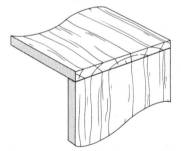

Q9 A hammer and another tool are used to drive the heads of bullet-head nails below the surface of timber. Name the other tool.

....

Q10 Why do we drive the heads of bullet-head nails below the surface of the timber?

....

....

....

Q11 What is the reason to use a nail with a large head as shown in the diagram at right? Record your reason.

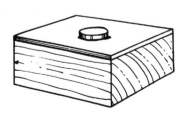

....

....

Nails

Nail types		Application
Bullet head nails or helical threaded nails Galvanised if corrosion resistance is required		**Hardwood and hardwood framing** **Edging and finishing work**
Flat head or threaded flat-head nails Galvanised if corrosion resistance is required		**Softwood and softwood framing**
Underlay nails Annular threaded		**Hardboard underlay sheets**
Wallboard nails Zinc plated		**Plywood sheets/panelling and** **wallboard for interior walls**
Hardboard nails Bright or zinc plated		**Hardboard**
Particle board nails Helical thread		**Particle board**
Flex sheet nails Galvanised		**Fibre cement sheets**
Decking nails Galvanised – helical thread		**Timber decking, timber cladding**
Clouts Galvanised		**Exterior thin sheet metal**
Bracket nails Galvanised		**Exterior thick sheet metal**
Roofing nail Plain galvanised		**Steel corrugated roofing**
Roofing nail Galvanised twisted shank		
Roofing nail Galvanised helical thread		
Sheet nails Zinc plated – annular thread		**Plasterboard sheets, e.g. Victorboard,** **Gyprock, Plasterboard**
Panel pins		**Finishing work and fine joinery**

Q12 There are many different types of nails for many different uses; see the table above. Through research or class discussion, list eight nail types and their uses.

1 ....

Use: ....

2 ....

Use: ....

3 ....

Use: ....

4 ....

Use: ....

5 ....

Use: ....

6 ....

Use: ....

7 ....

Use: ....

8 ....

Use: ....

Teacher's signature   Date

....

9780170439909 © 2019 Basil Slynko

Name.. Date..

Fasteners for metals

These activities will help you to:

▶ be able to identify different types of fasteners for metals

▶ gain knowledge of the different uses of rivets and screws

▶ know when to use different kinds of metal fasteners for different purposes.

A **fastening device** is one that holds two or more parts firmly together. Fasteners can be grouped into two basic categories:

- permanent fasteners
- semi-permanent fasteners.

When a **permanent fastener** is used, the parts that it joins cannot be separated without causing damage or destruction. There are two common types of permanent fastener:

- solid rivets
- blind (pop) rivets.

Solid rivets are named according to their head shape. They are used in general metalworking and often also to join plate material – that is, a thickness greater than 10 millimetres.

Q1 From the accompanying list of types of solid rivet, select the correct name for each of the rivets illustrated.

WORD BANK
Countersunk head
Flat head
Pan head
Round head
Truss head

_____ _____ _____ _____ _____

Q2 List three metals from which solid rivets are made.

1 .... 2 .... 3 ....

Blind (pop) rivets are used for speed and convenience, usually to join sheet metals. They are also used when the work cannot be supported from one side, or when it is accessible from one side only. Special pliers are used to draw the mandrel through the rivet.

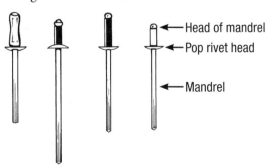

← Head of mandrel
← Pop rivet head
← Mandrel

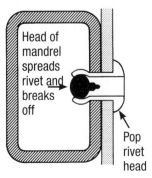

Head of mandrel spreads rivet and breaks off

Pop rivet head

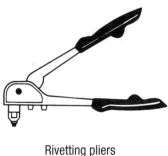

Rivetting pliers

Q3 List three reasons why blind rivets are generally used.

1 ....

2 ....

3 ....

To calculate how long a rivet should be when it is used to combine two sheets of metal, you should use the rule shown in this diagram. Note that the diameter of the rivet determines its strength.

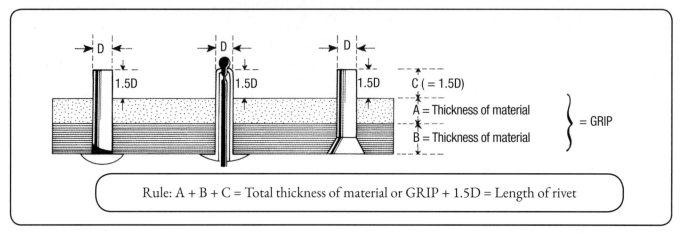

Rule: A + B + C = Total thickness of material or GRIP + 1.5D = Length of rivet

Q4 Circle which of these three rivet lengths will be needed to combine the two sheets of metal shown in the diagram at right.

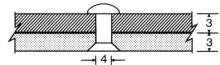

a 15 mm   b 12 mm   c 14 mm

Q5 You have just purchased a quantity of rivets 4 millimetres in diameter and 20 millimetres long. What is the maximum grip (material thickness) for which these rivets can be used? Circle the correct length.

a 12 mm   b 14 mm   c 15 mm   d 17 mm   e 20 mm

A **semi-permanent fastener** is used when it may be necessary to dismantle or adjust the components at a later date. The most common types of semi-permanent fastener are **threaded fasteners**:

• **bolts**: these have a solid head at one end and a metal thread at the other

• **nuts**: these are fitted to the ends of bolts and metal thread screws

• **screws**: either metal thread screws or self-tapping screws.

The threads on bolts, nuts and screws range from **coarse** thread or pitch to **fine** thread or pitch.

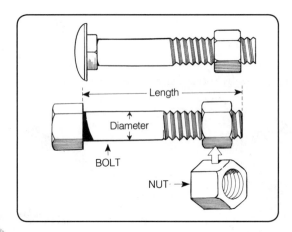

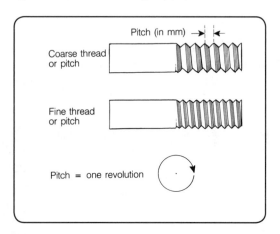

Q6 What are two geometric shapes commonly used for the heads of bolts?

1 ....   2 ....

Set screws are threaded along their shank as closely as possible to the head. They are used in threaded holes and therefore do not require nuts. Because they are generally used to join machine components, they are sometimes called **machine screws**.

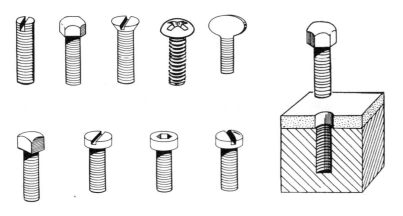

WORD BANK
Cheese head
Countersunk head
Grub screw
Hexagonal head
Raised head
Round head
Socket head
Square head
Thumb screw

Q7 In the above diagram, number any six screws with different head shapes and then use the word bank to name them here.

1 ....   4 ....   7 ....

2 ....   5 ....   8 ....

3 ....   6 ....   9 ....

Q8 In one or two sentences, explain why machine screws have such a wide range of head shapes.

....

....

....

....

....

Self-tapping screws cut their own thread path as they are screwed into a hole drilled in materials such as plastic, metal or wood, to secure one component to another. They are generally used to secure thin-gauge materials.

Self-drilling, **self-tapping screws** drill their own hole as well as cutting their own thread.

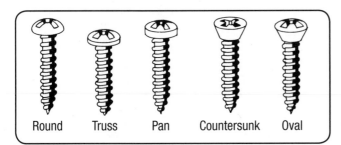

Round   Truss   Pan   Countersunk   Oval

Phillips self-tapping screws

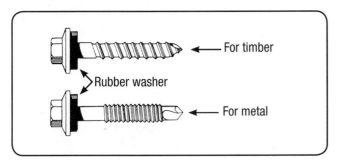

For timber

Rubber washer

For metal

Self-drilling, self-tapping roofing screws

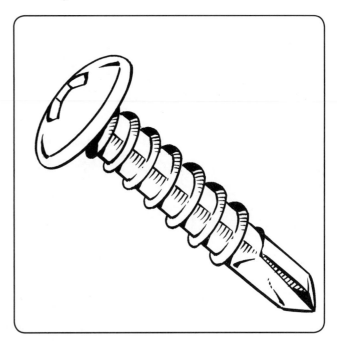

Pan head self-drilling, self-tapping screw

Q9 When ordering semi-permanent fasteners for metals, which factors should you bear in mind?

1 ....

2 ....

3 ....

4 ....

5 ....

6 ....

7 ....

 Always consider the **tensile strength** of bolts for each application. For example, bolts used for the construction of buildings would be high tensile, while a mild steel bolt would be used for the assembly of office furniture.

Q10 Draw lines to match each type of semi-permanent fastener with the situation in which it should be used.

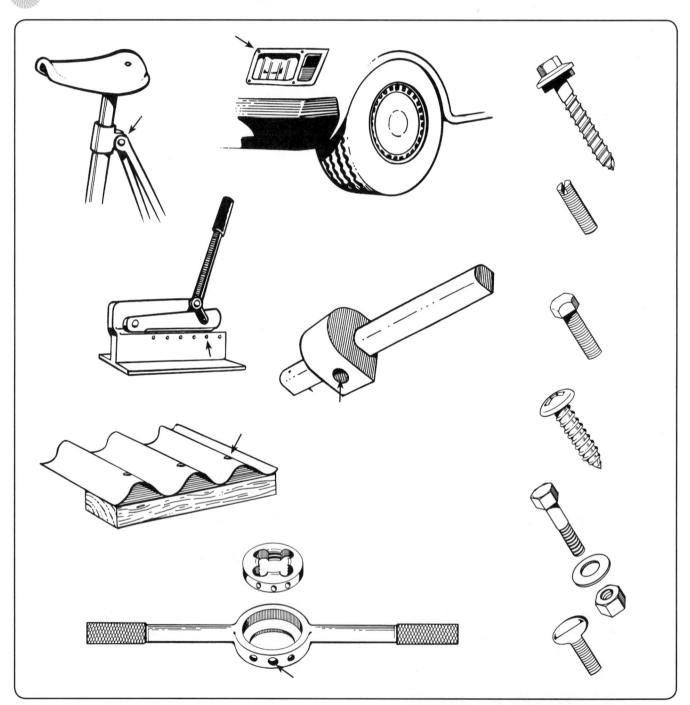

Teacher's signature   Date

....

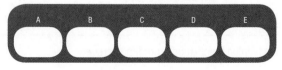

A   B   C   D   E

WORSHEET

Choosing adhesives

WORKSHEET

7.5

These activities will help you to:

▶ know for what uses particular adhesives are best suited

▶ be aware of how adhesives should be prepared and applied.

Most materials, whether they are similar or dissimilar, can be joined by either:

- pure animal or vegetable glues

  or

- polymer-based adhesives.

There are four main aspects to consider when selecting a glue or adhesive:

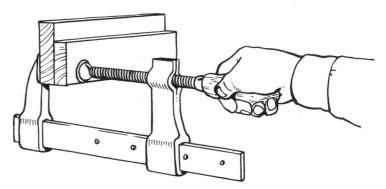

1 The **types of materials** to be joined

2 The **surface condition** of the materials

3 The kinds of **stresses** to which the join will be subjected

4 Whether the adhesion is required to be **permanent** or **temporary**.

Q1 Complete the table on the following page by inserting each of the factors below into one of the four sections of the table. (Some of the factors may apply to more than one section of the table.)

- Whether the materials are similar or dissimilar
- Whether the components may need to be dismantled in future
- Whether the surfaces must be free from grease and dust
- Whether the item is to be used indoors or outdoors
- Whether the surface must be slightly roughened
- Whether the surface must be dry
- Whether the materials are thick or thin
- Whether the join is load-bearing
- Whether gap-filling properties are required
- Whether the materials are porous or non-porous

1	Types of materials to be joined	
2	Surface conditions	
3	Stresses on join	
4	Whether adhesion must be permanent or temporary	

Adhesive **curing times** may be lengthened by low atmospheric temperatures when adhesives are used in cold climates.

 Many adhesives are highly flammable and some also give off toxic fumes. Keep them away from naked flames and work in a well-ventilated area. Always use the correct PPE (personal protective equipment).

Q2 Use the numbers from the mind map on adhesives to choose suitable adhesives for bonding these materials. (The first line has been completed as an example.)

	Wood (interior)	Rubber	Bricks/concrete	Plastics	Plasterboard	Paper/cardboard	Metal	Hardboard	Glass	Cork	Ceramics	Wood (exterior)
Wood (interior)	1	5	4	3	5	1	4	1	3	1	4	2
Rubber												
Plastics												
Metal												
Glass												
Ceramics												

Note: When similar materials are to be combined, there might be more suitable means of bonding them than using adhesives (for example, mortar can be used for bricks, and solvents can be used for plastics).

Adhesives

6 RESORCINOL FORMALDEHYDE ADHESIVE – thermosetting

- Highly resistant to moisture
- Has good gap-filling properties
- Excellent for boat-building and for outdoor furniture
- Stains wood
- *Shelf life:* 3–6 months

1 POLYVINYL ACETATE (PVA) ADHESIVE – thermoplastic

- A white plastic emulsion in ready-to-use form
- A strong adhesive that is suitable for porous surfaces only
- Not completely waterproof
- Suitable for use with interior woodwork
- Not recommended for joins under high permanent load as the adhesive creeps when subjected to a load
- *Shelf life:* indefinite if container is kept sealed

2 UREA FORMALDEHYDE ADHESIVE – thermosetting

- Available in liquid or powder form
- Suitable for use with hard, resinous timber
- Resists moisture better than PVA adhesive
- Has some gap-filling properties
- Produces a strong bond after one week
- *Shelf life:* 6–12 months

ADHESIVES

5 CONTACT OR RUBBER-BASED ADHESIVE

- Both surfaces must be coated with a thin film of adhesive, which has to be 'touch dry' before the components are joined
- Not suited to woodworking
- No clamping is required
- Produces a flexible joint
- Suitable for bonding dissimilar materials (for example, bonding laminated plastics, fabrics or vinyl to timber)
- *Shelf life:* one year if gelling has not occurred

4 EPOXY RESIN ADHESIVE – thermosetting

- May be **two-part** epoxy resin (equal parts of resin and hardener have to be mixed) or **single-part** epoxy resin (the components are pre-mixed and heat cures the adhesive)
- Has good gap-filling properties
- Gives an extremely strong bond, with no shrinkage
- Suitable for bonding metals, glass, wood, rubber, ceramics and concrete
- Water-resistant
- Expensive
- *Shelf life:* 6–12 months

3 CYANOACRYLATE ADHESIVE ('super glues')

- Gives an extremely strong bond
- Cures quickly – handling strength is achieved within seconds
- No clamping is required
- Only a thin film is applied
- Suitable for bonding metals, plastics, rubber, ceramics, wood, leather, paper and cork in any combination
- *Shelf life:* one year

DANGER: Extra care is required when using 'super glues' because of their powerful adhesive properties

Teacher's signature   Date

....

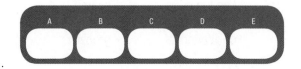

WORKSHEET
7.6

Methods of joining

These activities will help you to:

▶ gain knowledge of techniques used in joining materials

▶ know in which situations to use these techniques.

Materials can be combined using:

- mechanical fastening – using screws, nails, press-studs, zips and so on

- chemical adhesion – using glues and adhesives

- jointing – through interlocking of the components

- cohesion – by means of solvents or welding.

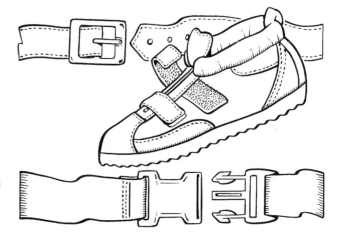

Pull-apart fasteners (sometimes called reclosable fasteners) such as zips, press-studs and velcro are generally used with natural materials such as textiles, rubber and leather. They allow for easy joining and separating of materials.

No tools are needed to open and close these fasteners. They use twin or mating parts that interlock.

Q1 Through research or class discussion, identify at least three other types of pull-apart fasteners that are used in your home, school or community, and state the purposes for which they are used.

Type 1 ....

Type 2 ....

Type 3 ....

Type 4 ....

Joints are an alternative to mechanical fasteners and adhesives when combining similar materials – for example, timber/timber, metal/metal, plastic/plastic and so on. Note: Fasteners and adhesives are often used with joints for added strength.

Q2 These drawings (below and on the following page) show some common joints for **widening** a range of materials. Use the word bank to name each joint shown. Also name at least one material for which each joint is commonly used.

 WORD BANK

Butt	Open seam
Dowelled butt	Rebate
Grooved seam	Run-and-fell
Lap (glued,	seam
soldered and/	Tongue and
or rivetted)	groove

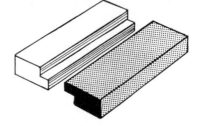

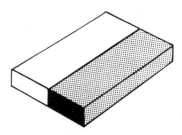

....

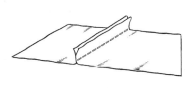

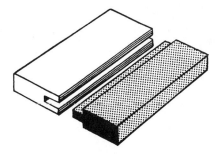

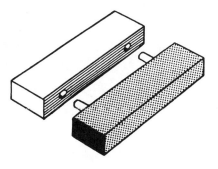

....

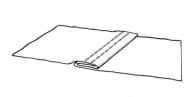

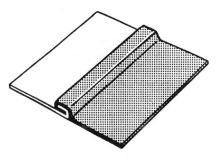

....

Q3 The biscuit shown on the right is used with which type of widening joint? (Refer to the word bank on the previous page.)

Biscuit

....

Q4 These drawings show some common joints used in **carcase construction** (that is, in box-like constructions). Use the word bank to name each joint shown.

> **WORD BANK**
> Box pin
> Butt
> Cleat
> Dovetail
> Knocked up
> Lap seam
> Mitred
> Peined down
> Rebated butt
> Stopped housing
> Through housing

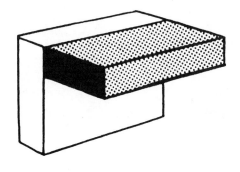

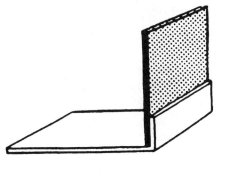

....

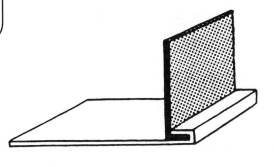

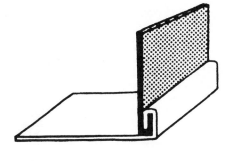

....

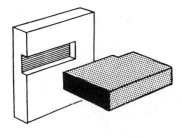
....

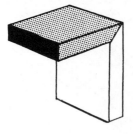
....

....

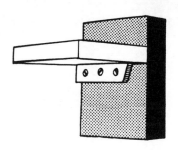
....

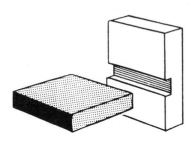
....

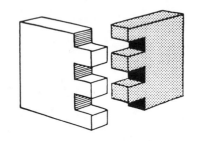
....

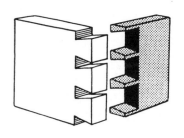
....

Q5 These pictorial drawings show some common joints used in **framing** (for example, in a window frame or house frame). Use the word bank to name each joint shown.

 WORD BANK
Bridle
Cleat
Corner halving
Cross halving
Dovetail halving
Dowelled butt
Gusset
Mitred
Mortice and tenon
Tee halving
Timber framing anchors

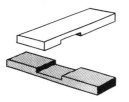
....

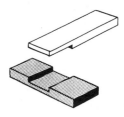
....

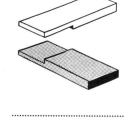
....

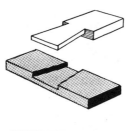
....

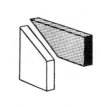
....

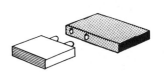
....

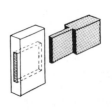
....

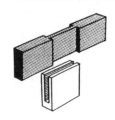
....

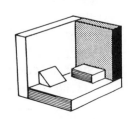
....

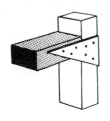
....

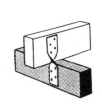
....

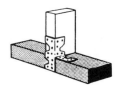

Most joints are considered to be permanent, but **knock-down (KD) fittings** are **semi-permanent**; they are used mainly to join manufactured boards where traditional jointing methods are not suitable. KD fittings are often used to assemble bulky items supplied in kit form such as flat-pack furniture.

Q6 Use the word bank to name each KD fitting shown.

WORD BANK
Bracket
Cam and bolt connector
Cross dowel and bolt connector
Plate
Two block connector

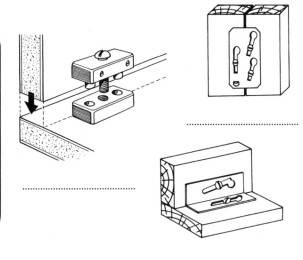

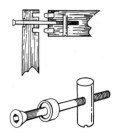

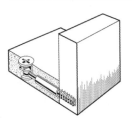

....

....

....

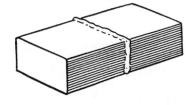

....

....

Q7 Write down one advantage and one disadvantage of knock-down fittings.

Advantage: ....

....

....

Disadvantage: ....

....

....

In the process of **cohesion**, similar materials are joined through the intermingling of molecules from adjacent surfaces. Cohesive bonding can be achieved by using **solvents** or by **welding**.

Q8 Which material(s) would be best suited to cohesive bonding?

....

Q9 Why would you use cohesive bonding in preference to other joining methods?

....

Q10 Give two examples of situations that would require cohesive bonding.

1 ....

2 ....

Teacher's signature   Date

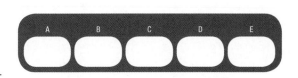

....

WORKSHEET

7.7 Finishing

These activities will help you to:

▶ gain knowledge of abrasives

▶ gain knowledge of finishes and finishing techniques.

Finishing is the process of preparing and coating the surfaces of any manufactured or constructed item. The function of finishing is to protect and beautify the surface.

Abrasives are particles of grit (natural or synthetic) that are bonded to a backing material – paper, cloth, foam – or bonded to one another in the form of a disc or wheel. Abrasives are used to cut away the top layers of material surfaces. An abrasive can be classified as either '**open coat**' or '**closed coat**'.

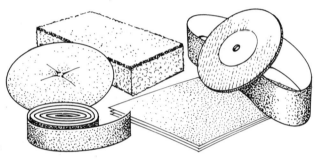

Q1 Unscramble the missing words in the following sentences and write them in the spaces provided.

PEON TOCA (........................ ....) abrasive papers are used on resinous timber that could quickly GCLO

(........................) the abrasive. SDEOLC ATOC (........................ ....) abrasive papers are generally used

on materials that OD TNO (........................ ....) readily OLGC (........................) the abrasive.

Abrasives are classified into grades according to the size of the mesh through which the grit particles pass. There are three basic kinds of abrasive: **coarse**, **medium** and **fine**.

Q2 Into which of the above three categories do these grades of abrasive fall?

a 200 to 600 ....   b 20 to 80 ....   c 100 to 200 ....

Q3 Render these drawings of pieces of timber to show the grain pattern. Draw arrows to show the direction in which the abrasive paper should be moved on each of the timber surfaces.

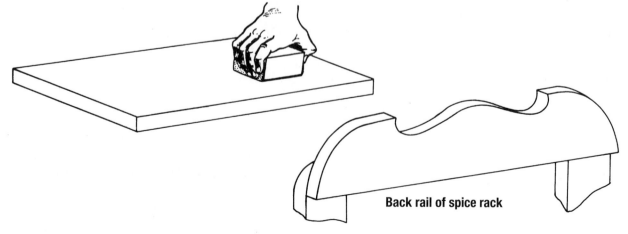

Back rail of spice rack

9780170439909 © 2019 Basil Slynko

Q4 A cork block should be used with abrasive sheets on flat surfaces. Give two reasons for this.

1 ....

2 ....

Q5 Use this word bank to complete the flow chart of the basic steps in finishing materials.

WORD BANK
Apply finish
Cleanse surface
Fill blemishes
Grind/sand – coarse
Grid/sand – fine
Plane/file
Scrape

Q6 Name two commercially finished sheetmetal products – other than tinplate and galvanised iron – that do not require additional finishing.

1 ....   2 ....

Paint is a general term for a preparation (something that is made) that acts as a preservative, protecting against moisture, chemicals and the sun's rays.

Q7 What are the three different coats that are usually applied when a bare surface is painted?

1 ....   2 ....   3 ....

Q8 Name the two general types of paint that are commonly used.

1 ....

Which cleaning agent is used for brushes and so on? ....

2 ....

Which cleaning agent is used for brushes and so on? ....

Transparent timber finishes are generally used to display or enhance the natural features of timber while still protecting it. They do not require special primers or undercoats.

Q9 List three types of transparent timber finish in general use.

1 ....   2 ....   3 ....

Staining is the technique of colouring wood. Stains do not conceal the grain of the timber.

Some common stains are:

• **water** stains – dissolved in water

• **spirit** stains – dissolved in methylated spirits

• **oil** stains – dissolved in turpentine.

Q10 Through research or class discussion, find and list four reasons for applying a stain.

1 ....

2 ....

3 ....

4 ....

Q11 This is a drawing of a piece of timber.

a Place a number in each circle to indicate the order in which you would apply a finish.

b Draw arrows to show where you would start to apply the finish, and the direction of the strokes on each surface.

Q12 Name three methods of applying finishes.

1 .... 2 .... 3 ....

Polishing is the term for a finishing process in which very small irregularities are removed from the surface of the material using an abrasive compound and a buffing cloth or mop.

Q13 Name one form of 'polish' suitable for use with each of these two materials.

Metal: ....

Acrylic: ....

Dyes are used to colour textiles and hides.

Q14 Name the two basic types of dye commonly used.

1 .... 2 ....

Q15 List two methods of dyeing fabric.

1 .... 2 ....

Inks can also be used to colour materials.

Q16 Complete the following sentence:

When using inks, special care is often required to prevent

....

Q17 The drawing above shows a range of other abrasives. Through research or class discussion, record the uses of each of the following abrasives.

1 Abrasive compound (e.g. Brasso): ....

2 Steel wool: ....

3 Grinding paste: ....

4 Buffing compound: ....

Teacher's signature   Date

....

A   B   C   D   E

WORKSHEET

8.1

Technology and the future

These activities will help you to:

▶ predict the possible effects of future changes in technology on people and the environment

▶ decide what your responsibilities are as a user and creator of technology.

Some technological solutions have proven to be unsatisfactory. The costs to people, society and the environment have been greater than the benefits. There is now greater awareness among people of the impact of technology on society and the environment.

So, as you look to the future, there are many questions to ask about the role of technology. Some of these are:

How will people use technology to solve their problems and meet their needs in the future? Technology will continue to shape our world, but what changes will it bring?	Will people develop new building materials and construction techniques so that they can create amazing new structures?	Will passengers be carried by faster and more efficient transportation systems?	New technology can improve people's lives, but as well as benefits there may be costs. We need to think about the possible harmful effects of new technology so that we can make wise decisions about using it.
		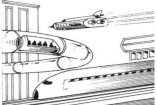	
Will users of technology have to conserve the world's resources of metals, timber and other materials more than they have in the past? Can we find new ways of re-using and recycling resources?	Will we learn to control the way we process materials so that the environment isn't harmed by pollution of the soil, water and atmosphere?	Will technology be used to help all the people of the world, or will it create a wider gap between the 'haves' and 'have nots' in different countries?	To make sure that technology is a good servant for everyone, we all need to be well informed about it, and we all need to have a say in how it is used to change the world.

Q1 As a member of a 'think tank', you have been given a list of some possible future developments in technology and asked to consider their effects.

Your task is to select from the list on the next page:

• one future development that you believe **will** benefit the world

• one future development that you believe **will not** benefit the world.

Then complete the details in the spaces provided on the next page. (You may need to carry out some research.)

In the future it is possible that:

- all petrol-powered vehicles will be banned from entering cities

- carbon dioxide emissions from the burning of fossil fuels will be captured and stored underground

- people will live and work in space

- buildings more than two kilometres high will be constructed

- automated highways will control traffic flow

- cities will be constructed under the sea

- hypersonic passenger aircraft will become common

- climate-controlled farming will increase crop production

- medical science will increase people's life spans.

Positive development: ....

Effects of technology	
Benefits	**Costs (if any)**
Summary of benefits versus costs	

Negative development: ....

Effects of technology	
Costs	**Benefits (if any)**
Summary of benefits versus costs	

Teacher's signature   Date

....

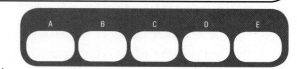

Name... Date...

The nature of technology

Some technologies have changed a great deal over the years, while others have not changed much. Every day, you use many of these technological solutions. It could be at home, at school or in the community – for example, a telephone, a computer or a calculator.

Yesterday   Today

Getty Images/E+/FG Trade

Challenge

Select a technological solution that interests you and complete a timeline below. Your task is to:

- trace its development
- show how the solution has changed
- record its impact on society and the environment

- note the use of new resources such as knowledge, materials and energy
- predict future trends and possible consequences.

My technological solution is ....

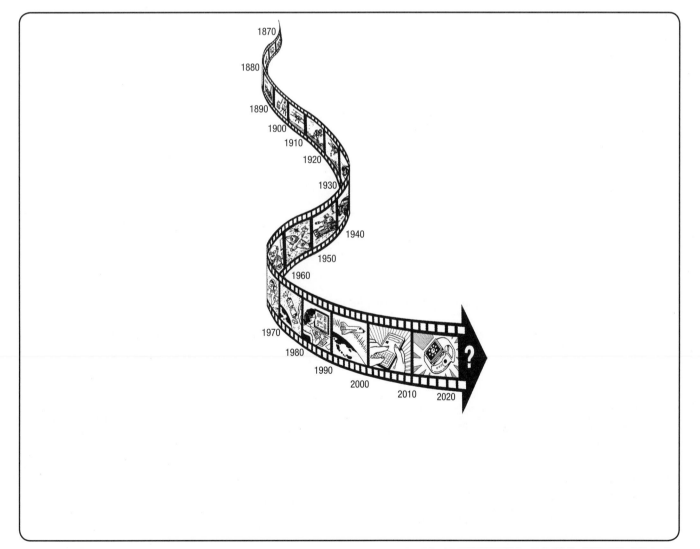

Teacher's signature   Date

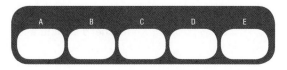

....

Name... Date...

Good design is usually the result of combining the elements and principles of design in an interesting or pleasing way – one that also takes into account factors such as safety, ergonomics and performance. This means that the designer has to be creative – that is, to have a good imagination.

Challenge

Q1 These two drawings show some shapes and lines. Complete each drawing to create a picture.

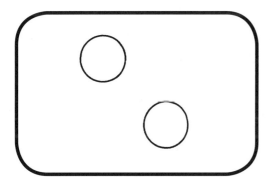 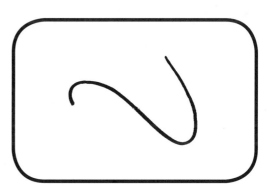

Q2 Look at this shape and record all the images that it brings to your mind, for example – a ball, a burnt bread roll, a hole …

....
....
....
....
....
....

Attach a sheet of paper to this page if you need more space for your answers.

Q3 When you **brainstorm**, you form a group and put everyone's ideas together. By brainstorming, how many different uses can you come up with for

- a used plastic shopping bag?

- these offcuts from a tree trunk?

Compare your group's answers with the answers of students in other groups.

Q4 **Group activity**

Using only one sheet of A4 paper for your group, build the tallest free-standing structure that is possible. (You may use glue, staples or 100 millimetres of adhesive tape.)

Teacher's signature   Date

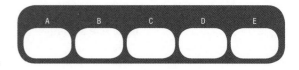

....

9780170439909 © 2019 Basil Slynko

Rendering

Challenge

A new model of a table jug is to be manufactured. You have been asked to prepare two advertising posters that feature rendered drawings.

With colouring pencils, felt-tip pens, watercolours or other appropriate media, render these two drawings.

Add foreground and background detail to the posters using various elements of design (see Worksheet 3.2) and principles of design. (The principles of design include balance, unity, proportion, contrast, emphasis, movement and rhythm.)

Poster A   Poster B

Q1 Which completed poster do you prefer? Why?

....
....
....
....

Teacher's signature   Date

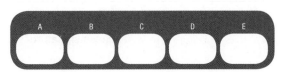

....

Name.. Date..

Special manufacturing aids have been developed to reduce the time that people take to complete particular tasks, or to reduce the likelihood that mistakes will be made. They include:

- **a template:** an outline that can be used over and over again to mark an accurate shape on material or to guide a tool when cutting the shape directly from the material.

- **a jig:** a device that guides a cutting tool into the right position for use. (It may also hold the material.)

- **a fixture:** a device that holds the material in place during processes such as drilling, sawing and assembly. Fixtures are usually attached to a surface or a machine.

Challenge

Your challenge is to develop manufacturing aids for the following tasks, and to sketch your solution(s).

Q1 Design a template to enable you to mark out the shape of the base of this peg board and the hole locations.

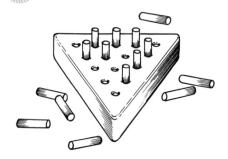
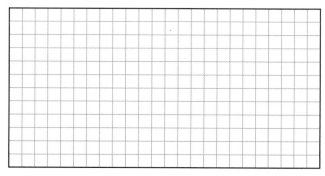

Q2 Design a jig to enable you to drill holes in the top of these salt and pepper shakers. (How can you hold the material?)

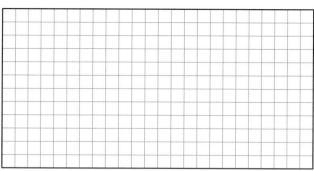

Q3 Design a fixture to enable you to bore a recess for the cork disc in the base of this drink coaster.

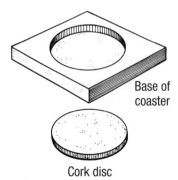

Base of coaster

Cork disc

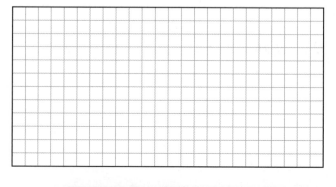

Teacher's signature   Date

....

Name.. Date...

Mechanism puzzles

Mechanisms are designed to perform specific tasks. They are often made up of many smaller mechanisms that work together to produce some kind of output motion from any input motion.

Challenge

Each of these 'black box' machines shows an input motion and an output motion. Your challenge is to design a mechanism that could be used to achieve the required output. The word bank lists some commonly used mechanisms.

WORD BANK

Cam	Lever
Chain and sprocket	Linkage
	Pulley
Crank	Ratchet
Crank and slider	Screw
Gear	

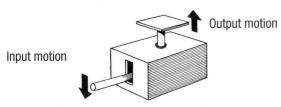

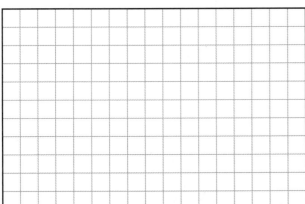

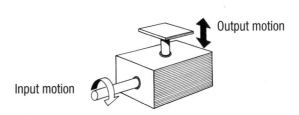

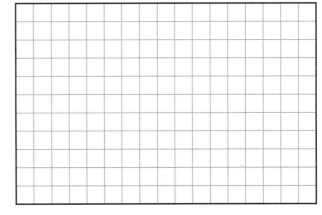

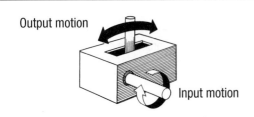

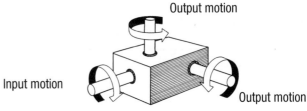

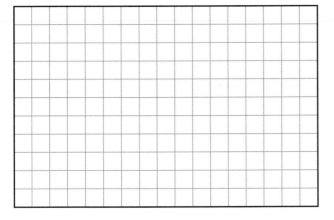

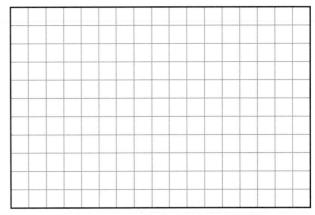

Teacher's signature   Date

....

Electronic circuit design

You have been given a schematic circuit diagram of an electronic doorbell. Your task is to design a circuit layout for making a printed circuit board (PCB) using the information in the component bank below, and not PCB software.

Remember: In a good circuit design, components are placed in rows, and similar components (such as resistors) are grouped together.

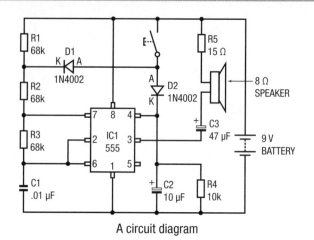

A circuit diagram

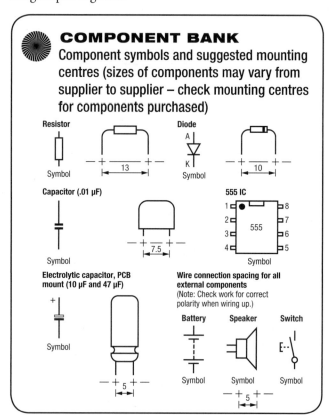

COMPONENT BANK

Component symbols and suggested mounting centres (sizes of components may vary from supplier to supplier – check mounting centres for components purchased)

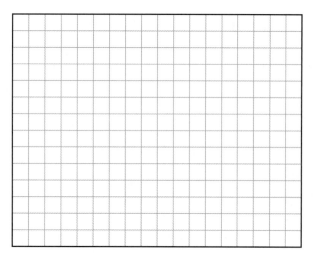

	DO	DON'T
Avoid using tracks that are the same size as the pads; otherwise the solder may flow away from the pads.		
Avoid sharp angles in tracks to prevent them from breaking.		
Always use the shortest possible track. This will minimise the resistance in the track.		
Maintain equal spacing where tracks pass between pads.		

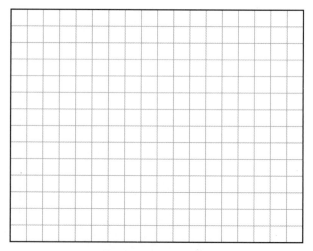

Some dos and don'ts of circuit layout

Component layout

Track layout (artwork)

    Teacher's signature   Date

....

9780170439909 © 2019 Basil Slynko

Name... Date.................................

Materials

People use some materials in their **natural** form. For example, tree trunks, stones and clay can be used as building materials. Other materials have to be **processed** before they can be used. For example, bauxite ore is refined to give the metal aluminium. There are also **synthetic** materials (such as plastics) that industrial chemists make from chemical compounds. Other materials are combined to produce **composite** materials such as Gore-Tex® - fabric laminated with plastic film and a thin layer of foam.

WORD BANK
Canvas
Cardboard
Concrete
Leather
Tempered glass
Woven cloth

Challenge

Select a material from the word bank (or choose another one that interests you). Then carry out research to answer the following questions.

The material I have selected is: ....

Q1 What is the raw material your selected material comes from?

....

....

....

Q2 How is the material processed? (You could draw a flow chart to answer this.)

Q3 What set of properties does the material have?

....

....

Q4 Where is the material used?

....

....

....

....

Teacher's signature   Date

....

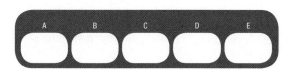

Name.. Date..

Personal safety

The aim of safety posters is to educate people to **think safety** and **act safely**.

When designing a safety poster, there are a number of factors that need to be considered. They are:

* The poster must be colourful and eye-catching.

* The poster must have a short message, to gain attention.

  If there is too much information, people are unlikely to read it.

* A picture can be used to emphasise the safety message.

* The message may be humorous.

* The poster should have a picture or diagram to support the message.

* The poster should be easy to read quickly.

Challenge

Design a safety poster for an activity that you enjoy, or for a work environment.

Teacher's signature   Date

....

9780170439909 © 2019 Basil Slynko

Portable power tools

Portable power tools (corded or cordless) can be used to increase the output of work in a given period. They can also improve the quality of work.

Most portable power tools are available in a range of sizes and have various capabilities (wattage, torque and rpm). Your choice of power tool depends on:

- the type of operation to be performed
- the length of time (and how often) the tool is used
- the type of material you are working with
- the cost of the tool.

Challenge

To complete this activity, you will need to refer to a power-tool catalogue, a range of power-tool brochures or visit an appropriate power-tool website.

Your friends want to build a dog house. They have asked you to provide sketches of dog-house design and construction details as well as technical advice on the selection of suitable power tools.

Your challenge is:

- to investigate dog-house designs and construction
- to identify the types of power tool required
- to compile a list of the most suitable models, with reason(s) for your recommendations. (You might also consider including a folio of pictures/brochures of the chosen power tools.)

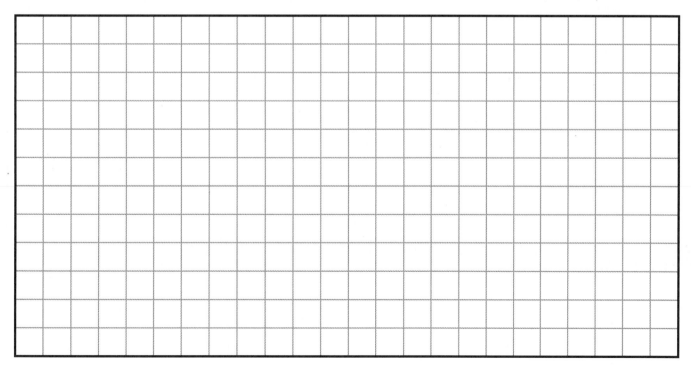

Sketch dog-house designs and construction details here

Name.. Date..

You could use the following sample format as a guide to select your power tools.

Operation	Power tool needed	Model and specifications	Reason(s)

Teacher's signature   Date

....